PHYSIOLOGICAL CHEMISTRY

A Series Prepared under the General Editorship of

Edward J. Masoro, Ph.D.

In preparation

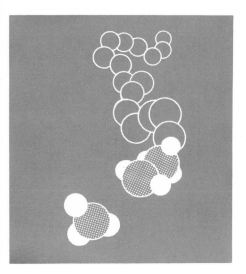

ENERGY
TRANSFORMATIONS
IN MAMMALS:
REGULATORY
MECHANISMS

Frederic L. Hoch, B.S., M.S., M.D.

*Associate Professor of Internal Medicine
and Biological Chemistry, The University
of Michigan Medical School*

1971

W. B. Saunders Company

Philadelphia · London · Toronto

W. B. Saunders Company: West Washington Square
Philadelphia, Pa. 19105

12 Dyott Street
London, WC1A 1DB

1835 Yonge Street
Toronto 7, Ontario

Energy Transformation in Mammals: Regulatory Mechanisms   SBN 0-7216-4700-6

Print No.:   9   8   7   6   5   4   3   2   1

TO J. L. ONCLEY

       . . . I this infer, —

That many things, having full reference
To one concent, may work contrariously:
As many arrows, loosed several ways,
Fly to one mark;
As many several ways meet in one town;
As many fresh streams run in one self sea;
As many lines close in the dial's centre;
So may a thousand actions, once afoot,
End in one purpose, and be all well borne
Without defeat. . . .

(Archbishop of Canterbury,
in *King Henry the Fifth,*
Act I, Scene 1, by
William Shakespeare.)

EDITOR'S FOREWORD

The past three decades or so have seen biochemistry emerge as possibly the most vigorous of the biological sciences. This, in turn, has led to a level of autonomy that has cut the cord linking biochemistry with its historically most important parent, mammalian physiology. For investigators in the fields of both biochemistry and physiology, this vitality has been most useful. But because of the arbitary separation of these two disciplines in most teaching programs and all textbooks, the vast majority of students do not see the intimate relationships between them. Consequently, the medical student and the beginning graduate student as well as the recently trained physician find it difficult, if not impossible, to utilize the principles of biochemistry as they apply to the physiological and pathological events they observe in man and other mammals.

Therefore, this series is designed not only to introduce the student to the fundamentals of biochemistry but also to show the student how these biochemical principles apply to various areas of mammalian physiology and pathology. It will consist of six monographs: (1) Physiological Chemistry of Lipids in Mammals; (2) Physiological Chemistry of Proteins and Nucleic Acids in Mammals; (3) Energy Transformations in Mammals; (4) Acid-Base Homeostasis: Its Physiology and Pathophysiology; (5) Physiological Chemistry of Carbohydrates in Mammals; and (6) Regulation of Amino-Acid Metabolism.

The series can be profitably used by undergraduate medical students. Recent medical graduates and physicians involved in areas of medicine related to metabolism should find that the series enables them to understand the theoretical basis for many of the problems they face in their daily work. Finally, the series should provide students in all areas of mammalian biology with a source of information on the biochemistry of the mammals that is not otherwise currently available in textbook form.

EDWARD J. MASORO

ACKNOWLEDGMENTS

I am grateful to a number of persons for their help in the completion of this manuscript; they are not responsible for its shortcomings, however. Drs. R. Estabrook and R. E. Beyer made suggestions after reading drafts. Dr. M. L. Ludwig revised the first chapters. I have profited by conversations with Drs. V. Massey, C. H. Williams, Jr., G. A. Palmer, and K. S. Henley. Mrs. F. Cullin has been patient in seeing the manuscript through its many revisions. A great number of investigators have evolved concepts that have become so much a part of one's thinking about bioenergetics and control mechanisms that they may not have been explicitly acknowledged here.

Ann Arbor, Michigan   FREDERIC L. HOCH, M.D.

CONTENTS

Part II PHYSIOLOGIC CONSIDERATIONS: ENERGY UTILIZATION

CHAPTER 15

CHAPTER 16

Part 1

Energetics and Oxidations

CHAPTER

1

ENERGY AND CARBON METABOLISM

Energy is defined as "the capacity to do work." A system of matter can be spoken of as "containing" energy, and the energy content is a property of that system. There are various forms of energy, e.g., thermal, radiant, potential, kinetic, and chemical. All of them can do work, i.e., exert a force through a given distance. The energy content of a system is diminished when work is done by the system on its surroundings, but in such exchanges energy is neither created nor destroyed, though it may be converted into a different form. A familiar example is the conversion of chemical to mechanical energy effected by the internal combustion engine.

Living organisms (it is almost a truism to point out) are composed of organized matter that is a "system"; as such they obey the primary laws of matter, and one of their properties is their content of energy. Rubner in 1891 first showed that the law of conservation of energy held for physiologic processes in mammals.

This book aims at describing the transformations of energy, or energy metabolism, in mammals and some of the factors that control energy metabolism. Several kinds of energy transformation occur in biologic systems.

Graham Lusk's definition that "metabolism is the flame of life" emphasizes the intracellular processes that convert chemical energy to heat. Besides heat production there are other processes that utilize the chemical energy of substances ingested by the living cell, e.g., reactions that transfer the chemical energy from one molecule to another, and those that utilize the transformed energy to do the mechanical work useful to maintaining life, such as muscle contraction.

The two fundamental laws of thermodynamics describe how *all matter* in the universe behaves. The First Law says that "the total energy of the universe is conserved, or stays constant." The Second Law says that "when any actual process occurs it is impossible to invent a means of restoring *every* system concerned to its original condition" (Lewis and Randall, 1923). The degree of irreversibility of real processes is defined as *entropy*, and the entropy of the universe increases. In colloquial terms, the First Law has been stated as "You can't win," meaning that you cannot increase the energy content of a system without borrowing the energy from another system. In these terms, the Second Law becomes "You can't even break even."

To discuss these laws, a "system" must first be defined: it is whatever part of the objective world that is chosen as the subject of thermodynamic discourse. A biologic system contains matter distributed heterogeneously within its boundaries. Once the boundaries are defined, the energy content can be described by measuring selected gross macroscopic properties of the system, such as its volume, temperature, and pressure. The total energy content of a material system may be enormous. In biologic systems, not the total energy content ($E = mC^2$)—we are not directly concerned here with hydrogen bombs—but the relatively much smaller increases and decreases of the energy content accompanying certain chemical reactions—the *changes* in energy—are of interest. To measure energy changes it is not necessary to know the total energy of the system in its initial and its final state, but only the energy the system has taken up from or given off to its surroundings, in passing from its initial state to its final state.

The change in the internal energy (ΔE) of any system, defined in the First Law in terms of Q (heat) and W (work):

Equation 1–1 $$\Delta E = Q - W$$

is the difference between the energy put into the system by the environment and the work performed by the system on the environment, and is given by convention a positive or negative sign. When the work done is smaller than the energy input Q, the energy change is negative; a negative energy change means that the system in its initial state had a higher energy content than in its final state. If the system is purely a mechanical one, and there is no change in temperature, no heat is transferred and the change in internal energy equals the work done by the system: $\Delta E = -W$.

Another quantity, the change in *enthalpy*, $\Delta H = \Delta E + P \Delta V$, is frequently used in applying the First Law. When the volume of a system is held constant, and only expansion work considered, then $\Delta H = \Delta E + Q$. It is often possible to study reactions at constant volume, and thus to determine ΔH by measuring the heat evolved or taken up. For instance, when glucose is burned completely (i.e., combines with oxygen according to Eq. 1–2) in a calorimeter, a measurable amount of heat is given off:

Equation 1–2 $\qquad C_6H_{12}O_6 + 6\ O_2 \rightarrow 6\ CO_2 + 6\ H_2O$

When one mole of glucose (180.16 g) is combusted, $\Delta H = -673$ kcal, where kcal is the kilocalorie, defined as that amount of energy necessary for raising the temperature of one kilogram of water from 14.5° to 15.5°. Thus, glucose plus oxygen has a higher energy content than the products, $CO_2 + H_2O$.

The change in energy content when glucose is oxidized by the cell is not all available for useful work, however. To measure the utilizable energy, a new parameter, the change in *free energy*, ΔG, must be considered.

The definition of free energy involves the Second Law:

Equation 1–3 $\qquad\qquad \Delta G = \Delta H - T\ \Delta S$

where ΔG is the change in free energy (also termed ΔF in older texts and in many tabulations), T is the absolute temperature and ΔS is the change in entropy. The units of ΔG are calories per mole, and the units of ΔS are calories per mole-degree. Equation 1–3 implies that the sign and the magnitude of ΔG depend upon both the change in the internal energy of the system (in chemical reactions in living systems, $\Delta H \simeq \Delta E$) and the change in entropy or randomness. ΔG for the oxidation of glucose is -686 kcal per mole. The large negative free energy changes accompanying the oxidation of carbohydrates, fats, and proteins provide the energy continuously required by mammalian systems.

The free energy change of a chemical transformation can be calculated from a variety of measurements: equilibrium constants, redox potentials, enthalpy and entropy changes, and free energies of formation. In a chemical system at equilibrium, no further net chemical change occurs, and no further net exchange of energy with the environment. The equilibrium constant, K_e, expresses the relationship between the thermodynamic activities ($\simeq$ mass concentrations) of reactants and products under those conditions. In the reaction,

Equation 1–4 $\qquad\qquad A + B \rightleftarrows C + D$

$\qquad\qquad\qquad\qquad$ (reactants) $\quad$ (products)

at equilibrium,

Equation 1–5 $\qquad K_e = \dfrac{(C)\quad (D)}{(A)\quad (B)} = \dfrac{(\text{products})}{(\text{reactants})}$

where the parentheses indicate concentrations in units of moles per liter. The equilibrium constant is a function of the change in free energy of the components of the reaction:

Equation 1–6 $\Delta G^{\circ} = -RT \ln K_e$

where R = gas constant (1.987 cal mole^{-1} degree^{-1}); ln K_e = natural logarithm of the measured equilibrium constant; and ΔG° is the *standard* free energy change defined under conditions in which the concentrations of the reactants are 1 molal (in reactions involving H$^+$ ions as reactants or products, the standard concentration is 10^{-7} M, not 1 M; the change in free energy at pH 7.0 is termed $\Delta G'$). ΔG° for glucose is -686 kcal per mole, higher (more negative) than the heat content (-673 kcal per mole) by the amount $T\Delta S$ (as indicated in Eq. 1–3); at $T = 298^{\circ}$, $\Delta S = -43.6$ cal per mole. Klotz (1967) stresses the subtle point that the free energy change calculated from measurements of K_e is in general *not* for the transformation at equilibrium concentrations (ΔG is always zero for a reaction at equilibrium), but for a *non*equilibrium standard state in which all concentrations (or activities) are unity.

   Few biochemical reactions occur under standard conditions. The value for ΔG will not equal ΔG° but will depend upon the actual concentrations of the reactants. An example (from Klotz, 1967) is the hydrolysis of adenosine triphosphate (ATP^{4-}). Under standard conditions, the starting concentration of all reactants or products is 1 M: ATP^{4-} = 1 M, or ADP^{3-} = HPO$_4{}^{2-}$ = 1 M.

Equation 1–7 $ATP^{4-} + H_2O \rightarrow ADP^{3-} + HPO_4{}^{2-} + H^+$

$$\Delta G^{\circ} = -7 \text{ kcal/mole}$$

The change in chemical potential is -7 kcal per mole. When the reactant and product concentrations are not standard and the pH is 7, and we abbreviate HPO$_4{}^{2-}$ to P$_i$:

Equation 1–8 $\Delta G = G^{\circ} + RT \ln \dfrac{ADP \times P_i}{ATP}$

ΔG can be calculated from the value of ΔG° by making assumptions about the concentrations of ADP, ATP, and P$_i$ in tissues. For example, let the concentrations of ADP and ATP be equal and P$_i$ = 1 mM (these values are in reasonable accord with chemical measurements). The value of ln (ADP $\times$ P$_i$/ATP) in Equation 1–8 will be that of ln P$_i$ = -6.9, and $RT \ln$ P$_i$ = -4300. Under these conditions, $\Delta G = -11.3$ kcal per mole, as compared with $\Delta G^{\circ} = -7$ kcal per mole. Thus, if in tissues the ADP, ATP, and HPO$_4{}^{2-}$ concentrations are maintained at some nonequilibrium (steady-state) values, the yield of free energy will be quite different from the standard

Table 1-1   *Free Energy Changes of Chemical Reactions**

		$\Delta G'$†		
		kcal	kcal	kcal
		mole	carbon	g
Oxidation	glucose + 6 O_{2g} → 6 CO_{2g} + 6 H_2O	−686	−114	3.74
	lactate + 3 O_{2g} → 3 CO_{2g} + 3 H_2O	−326	−108	3.62
	palmitate + 23 O_{2g} → 16 CO_{2g} + 16 H_2O	−2338	−146	9.30
	glycine + 3 O_{2g} → 3 CO_{2g} + NH_3 + 2 H_2O	−234	−117	3.12
Hydrolysis	sucrose + H_2O → glucose + fructose	−5.5		
	glycylglycine + H_2O → 2 glycine	−4.6		
Rearrangement	glucose-1-phosphate → glucose-6-phosphate	−1.745		
Ionization	AcCOOH → H^+ + Ac^-	+6.310		
Elimination	malate → fumarate + H_2O	+0.750		

* Values taken in part from Lehninger, 1965.
† ΔG is given as $\Delta G'$, the standard free energy change at pH 7.0, because under biologic conditions neutrality is usually assumed.

free energy, $\Delta G°$. At lower concentrations of free P_i, ΔG will become even greater (i.e., even more negative).

The amounts of energy liberated, i.e., the free energy of a reaction, of various chemical processes can be compared to show why mammals oxidize for maintenance of life, rather than obtain their energy by other chemical means. The oxidation of carbohydrates, fats, and amino acids to CO_2 and water involves a free energy change several orders of magnitude higher than that of the other kinds of chemical reactions listed in Table 1-1. Thus oxidation is a more efficient and economical process for the liberation of energy than the other processes in terms of the energy output per amount of fuel reacting. The size of the change of oxidative free energy, when these organic molecules react with oxygen to produce CO_2 and H_2O, depends upon the initial state of reduction of the carbon atoms. In fats, which contain many —CH_2—CH_2— groups, the carbons are highly reduced, and $\Delta G'$ is almost −150 kcal per g atom of carbon. In carbohydrates, with the characteristic —HCOH—HCOH— grouping, the carbons are already partly oxidized, and $\Delta G'$ is about −110 kcal per carbon. The oxidation of one gram of fat thus liberates more energy than the oxidation of one gram of carbohydrate. As will be seen in Chapter 7, the caloric equivalents of nutrients are calculated from the heat production resulting from the oxidative catabolism of fats, proteins, and carbohydrates, using values for the respiratory exchange of whole animals.

In biologic systems, the overall oxidative reactions, with their large free energy change, occur not in a single step but as a series of reactions with smaller changes in free energy that lead to the same products. The stepwise reactions are, at some points, coupled with processes that result in the transformation of the liberated energy into another chemical form—an energy-rich bond—or into a physical form—an ion-gradient potential or heat. The details of these transformations of energy will occupy the next chapters. At this point, we may consider in general terms the nature of energy-rich chemical bonds and the steps that are involved in the energy metabolism of the carbon atoms in organic substrates.

THE "HIGH-ENERGY BOND"

In biologic systems, the energy liberated through oxidative steps is *transferred* in part to the so-called "high-energy bonds" of various compounds. A high-energy bond is one that reacts at neutral pH with a concomitant large negative free energy change. The reaction may involve a transfer of electrons or a hydrolysis, and the free energy change is more negative than −7 kcal per mole. The substances that react with "high-energy bonds" are usually present in high concentrations; for instance, water is easily available in biologic systems. In a sense, then, water can be considered just as much a "high-energy compound" as ATP, since it is the interaction between the two that is accompanied by such a large free energy change.

Although the concept of high-energy bonds has been highly productive in biochemistry, the term itself is misleading and has been objected to by many observers, in particular, chemists. The energy from oxidations is not "stored" in the terminal pyrophosphate bonds of ATP; there are "changes in chemical potential, ΔG, when certain groups are transferred from one molecule to another; a more appropriate name might be group-transfer potential" (Klotz, 1967). Several types of transfer potentials are important in biologic reactions: *acidity*, where the potential is proportional to the $\Delta G°$ per mole of H^+ transferred; *electron-transfer potential*, proportional to the $\Delta G°$ per mole of electrons transferred; and *group-transfer* potential, proportional to the $\Delta G°$ per mole of group transferred.

The phosphate-transfer potential is high when certain molecules found in biologic systems are hydrolyzed or transferred to other molecules, and these are termed "high-energy phosphate" bonds. Phosphoric acid anhydrides, phosphoric-carboxylic anhydrides, phosphoguanidines, and enol-phosphates (Fig. 1–1) all react with water or other molecules with an accompanying free energy change equal to or greater than −7 kcal per mole. Adenosine triphosphate (Fig. 1–1) is a phosphoric acid anhydride. The hydrolysis of ATP at pH 7, which proceeds according to Equation 1–7, has a large negative free energy change because the phosphoric anhydrides are so much less stable than their products. These differences arise from the strong electron-withdrawing property of the phosphate group, and lesser resonance energy and greater electrostatic repulsions of the anhydride trianion and tetra-anion molecule as compared with those of the two phosphate anions that are produced (Mahler and Cordes, 1966). In addition, the importance of the liberation of the H^+ ion (Eq. 1–7) in the hydrolysis of ATP has been stressed (George and Rutman, 1960).

These chemical and physical descriptions of high-energy compounds and their hydrolytic products are specific ways of stating that the reactants are at a higher energy state than the products—a necessity for reactions proceeding with a negative free energy change. The specific features of group-transfer and other transfer reactions make more understandable the observations that not all phosphate-transfers, for instance, involve free energy changes greater than -7 kcal per mole: the hydrolysis of glucose-6-phosphate is accompanied by a $\Delta G° = -3$ kcal per mole. The products of this hydrolysis are at energy levels not so far below that of the phosphate ester as are the products of the hydrolysis of the phosphoric anhydrides. The existence of relatively large changes in free energy during chemical or enzymatic hydrolysis should not be taken to mean that ATP and other compounds possessing an energy-rich group are unstable and decompose spontaneously in solution; ATP is quite stable.

High-energy phosphate bonds are found not solely in the two pyrophosphate linkages of ATP. In mammalian muscle, the main storage form of high-energy is in the phosphoamide bond of creatine phosphate (in lower animals, in arginine phosphate). Other forms of high-energy phosphate bonds are phospho-enol-pyruvate, acetyl phosphate, pyrophosphate, and polypyrophosphate. In addition, there are other high-energy bonds that do not involve phosphate; other chemical groups that have a high transfer potential are thioesters like acetyl coenzyme A, sulfonium compounds like S-adenosylmethionine, and amino acyl esters of the ribose moiety of nucleic acids (see Fig. 1–1). High-energy bonds are often represented symbolically by a "squiggle," $\sim$, e.g., $\sim$P. Thus, the energy-rich bonds of ATP are represented as adenosine-P$\sim$P$\sim$P, and creatine phosphate as creatine $\sim$P; however, acetyl coenzyme A is usually not provided to the reader with a squiggle.

A feature of the biologic compounds that contain $\sim$P-bonds is their ability to transfer phosphate groups readily among themselves. The presence of many *kinases*, enzymes that catalyze such transfers, and the very small change in free energy that accompanies the transfer to phosphate groups from one of these substances to another, facilitate the interchanges. Indeed, the storage and the utilization of energy in the form of high-energy chemical bonds depend upon these transfers of $\sim$P-bonds and other $\sim$-bonds. An example is the exchange between ATP and creatine $\sim$P,

Equation 1–9

Adenosine-P$\sim$P$\sim$P + Creatine $\rightleftarrows$ Adenosine-P$\sim$P + Creatine $\sim$P,

catalyzed by the enzyme creatine phosphokinase. The standard free energy change during the hydrolysis of creatine $\sim$P at pH 7.7 is -9 kcal per mole; the $\Delta G°$ that occurs when ATP phosphorylates creatine is thus $+2$ kcal per mole, and a slight tendency exists for creatine $\sim$P to phosphorylate ADP. Other energy-rich bond exchanges occur with even smaller free energy

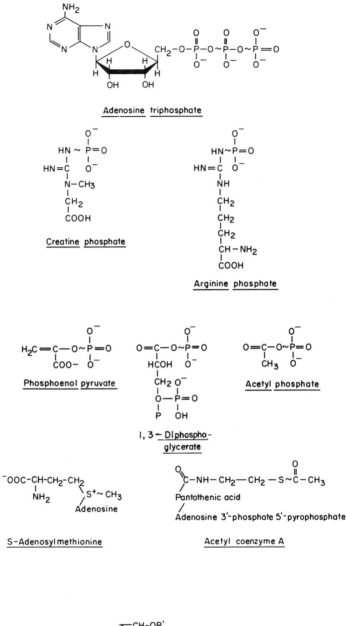

Adenosine triphosphate

Creatine phosphate

Arginine phosphate

Phosphoenol pyruvate

1, 3 — Diphospho-
glycerate

Acetyl phosphate

S—Adenosyl methionine

Pantothenic acid

Adenosine 3'-phosphate 5'-pyrophosphate

Acetyl coenzyme A

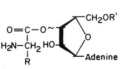

Amino acyl – tRNA

Figure 1–1. Some biologically important molecules that contain "high-energy" bonds.

changes; the $\Delta G°$ of thiol ester hydrolysis is -7.7 kcal per mole. With such small free energy differences between the phosphate-transfer potentials of ATP and the other energy-rich bonds, these reactions require little or no energy input; they are iso-energetic, and there is no energy barrier to their proceeding.

CARBON METABOLISM

Figure 1–2 diagrams some of the major pathways whereby the carbon atoms of substrates are oxidized to CO_2 and H_2O, and the energy of oxidation is transformed to $\sim$P-bonds. A detailed description of the many reactions involved in these oxidations and interconversions is beyond the purposes of this text, and the reader is referred to the other volumes in this series that deal with lipids (Masoro, 1968), proteins (Kaldor, 1969), and carbohydrates (Shreeve, in press), as well as to standard textbooks of biochemistry.

The oxidative metabolism of fats, carbohydrates, and proteins is shown in Figure 1–2 as falling into three phases according to the degree of oxidation, after the original outline of Krebs and Kornberg (1957).

In Phase I, the large polymeric molecules of the food are hydrolyzed into monomeric components: proteins to amino acids, carbohydrates to hexoses, and fats to fatty acids and glycerol. Each bond that is broken in this phase liberates relatively small amounts of energy, between 2.5 and 4.3 kcal per mole, as expected from hydrolyses. Only 0.6 per cent of the free energy of oxidation of proteins and polysaccharides, and 0.1 per cent of that of fats, is liberated. Phase I prepares the foodstuffs for major energy transfers and forms no high-energy compounds.

In Phase II, incomplete oxidations of the monomeric units transform the large number of diverse molecules into one of three substances—acetyl coenzyme A, α-ketoglutarate, or oxaloacetate—that are metabolized further in the next phase. Acetyl coenzyme A constitutes the greatest amount, representing two-thirds of the C of carbohydrates and glycerol, all the C of the common fatty acids, and about one-half the C of the amino acids. Of the total free energy of oxidation, about one-third is transformed in Phase II. Besides decreasing the diversity of organic compounds from a very large number to only three that can subsequently enter a restricted common oxidative pathway, Phase II also produces energy-rich P-bonds (their fate and importance will be discussed presently) and CO_2. The glycolytic pathway, whereby a 6-carbon hexose is anaerobically hydrolyzed and oxidized to two 3-carbon pyruvate molecules, involves as well the esterification of P_i, and the evolution of two $\sim$P-bonds per hexose.

In Phase III, the 2-carbon (acetyl coenzyme A), 4-carbon (oxaloacetate), and 5-carbon (α-ketoglutarate) products of metabolism enter the pathway of the *tricarboxylic acid cycle*. The products, when each molecule of acetyl coenzyme A is metabolized, are two CO_2 molecules, a single $\sim$P-bond

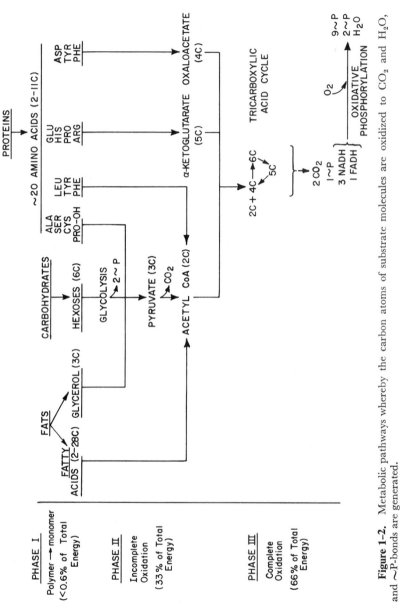

Figure 1-2. Metabolic pathways whereby the carbon atoms of substrate molecules are oxidized to CO_2 and H_2O, and $\sim$P-bonds are generated.

(formed anaerobically by a process analogous to those in glycolysis), and four molecules of reduced coenzymes (three of reduced nicotinamide adenine dinucleotide, and one of reduced flavin adenine dinucleotide). These coenzymes are discussed in the next chapters; it is enough to point out here that they both have a high electron-transfer potential (although they are not usually thought of as "high-energy" compounds) and that their oxidation involves the uptake of O_2 (respiration) and the formation of H_2O and 11 $\sim$P-bonds via the process of *oxidative phosphorylation*, which accounts for the release of the remaining two-thirds of the free energy of oxidation of fats, proteins, and carbohydrates. In Phase III, part of the free energy of oxidation of the three organic products of Phase II is transferred to the two coenzymes, and they in their turn transfer a part of their free energy of oxidation to form $\sim$P-bonds.

General References: The reader is referred to standard textbooks of thermodynamics for rigorous derivations and definitions. A classic text is Lewis and Randall (1923). For a more discursive but terse discussion that is pertinent to biochemistry, and pleasurable, see Klotz (1967). Linford (1966) develops the concepts in words, but less tersely. Lehninger (1965) outlines the basic thermodynamics and the biochemistry. Lipmann (1941) discusses the concepts of metabolic generation of phosphate bond energy, and George and Rutman (1960) write about the high-energy bond.

2

ENZYMATIC DEHYDROGENATION: SUBSTRATE-LEVEL PHOSPHORYLATION

Stepwise enzymatic oxidations liberate the energy of ingested large organic molecules in biologic systems. The enzymes and cofactors involved are very similar in the different tissues among the mammalian species, and are found as well in the lower animals and organisms. In general, the first oxidative step utilizes an enzyme that specifically oxidizes a certain substrate molecule; there are many different substrates, and there are many different enzymes that are specific for one or more substrates. These enzymes are dehydrogenases that remove a pair of hydrogen atoms and electrons from the substrate.

The dehydrogenases are proteins that act in conjunction with small organic molecules (coenzymes). The coenzyme itself participates in the hydrogen-transfer and in the electron-transfer, and the same coenzyme is the necessary partner for a number of different enzymes that catalyze the dehydrogenation of different molecules. The enzyme itself imparts the specificity for the type of substrate that is attacked, combining with the substrate in a special manner.

In mammalian (and other) systems, the initial dehydrogenation of a substrate can be classified according to the organic group that is oxidized

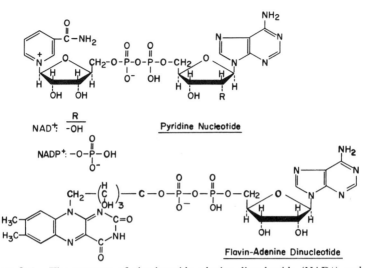

Figure 2–1. The structure of nicotinamide adenine dinucleotide (NAD⁺) and flavin-adenine dinucleotide (FAD).

or the coenzyme that is involved. These two classifications are inter-dependent; i.e., the coenzymes have a general specificity for an organic group. There are two types of coenzymes for substrate dehydrogenases: the pyridine nucleotides and the flavins. Their structure in the oxidized state is shown in Figure 2–1.

PYRIDINE NUCLEOTIDES

The *pyridine nucleotide-dependent dehydrogenases* catalyze the oxidation of a C—H bond in one of several organic groups, depending upon the specificity of the protein apoenzyme (Fig. 2–2).

The variety of these groups speaks for the widespread role of the pyridine

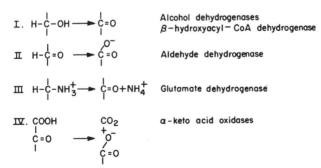

Figure 2–2. Typical oxidations catalyzed by pyridine nucleotide-dependent dehydro-genases.

nucleotides in substrate dehydrogenations, especially in the conversion of alcohol and amino compounds to carbonyl compounds and of carbonyl compounds to carboxylic acids. This role becomes even more significant when it is considered that the L-amino acids are not oxidized directly to any significant degree in mammalian systems but are first transaminated to α-keto acids and then oxidized via pyridine nucleotides.

The first step in the participation of the pyridine nucleotides in enzymatic dehydrogenations is the attachment of the oxidized pyridine nucleotide to the enzyme (E). The binding is strong enough that the enzyme-coenzyme complex sediments in a gravitational field and can be isolated thereby, but it is not so strong that the complex cannot dissociate readily upon dilution.

Equation 2–1 $E + NAD^+ \rightleftarrows E \cdot NAD^+$

The enzyme-coenzyme complex, $E \cdot NAD^+$, is catalytically active; neither the free enzyme nor the free coenzyme can itself dehydrogenate the substrate. Each dehydrogenase binds two or more molecules of a coenzyme and each site of coenzyme binding acts as an independent catalytic locus. In several, but not all, the dehydrogenases, an atom of zinc is involved in the binding of the coenzyme. The strength of the binding of the oxidized coenzyme, as expressed by a binding constant of 10^4 to 10^6 M, is about 10 to 100 times weaker than that of the reduced coenzyme ($K = 10^5$ to 10^8 M). The pyridine nucleotide coenzymes are thus relatively easily dissociable from their apoenzymes and are more like cosubstrates (the substrates bind to the enzyme-coenzyme complexes with a $K = 10^3$ to 10^5 M). This ease of dissociation is economically advantageous, allowing one molecule of coenzyme to serve many molecules of apoenzyme.

The second step is the combination of the enzyme-oxidized-pyridine-nucleotide complex with a substrate (SH_2) molecule.

Equation 2–2 $E \cdot NAD^+ + SH_2 \rightarrow E \cdot NAD^+ \cdot SH_2$

These enzyme-coenzyme-substrate complexes are easily dissociable and are very transient. That they do exist is proved by the fact that a transfer of hydrogen occurs between the substrate and the coenzyme in the third step of the catalytic mechanism.

Equation 2–3 $E \cdot NAD^+ \cdot SH_2 \rightarrow E \cdot NADH + S + H^+$

The hydrogen transfer between SH_2 and $E \cdot NAD^+$ is stereospecific and involves definite hydrogen atoms on the substrate and on the coenzyme. An example is the action of alcohol dehydrogenase in oxidizing ethanol to acetaldehyde (Fig. 2–3).

The transfer of the hydrogen atom in this case is called A-stereospecific;

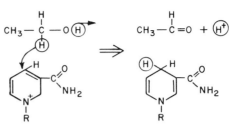

Figure 2–3. The stereospecific transfer of an H atom, catalyzed by the pyridine nucleotide-dependent alcohol dehydrogenase.

other dehydrogenases transfer the substrate H to the other position on the C_4 of the nicotinamide and are called B-stereospecific. There does not seem to be any underlying correlation between the stereospecificity and the nature of the reactions.

The fourth and final step in catalysis is the dissociation of the enzyme-reduced-pyridine-nucleotide complex to liberate the free enzyme and NADH:

Equation 2–4 $E \cdot NADH \rightarrow E + NADH$

The sequence of steps in this catalytic dehydrogenation can be summarized as:

Equation 2–5 $SH_2 + NAD^+ \rightleftarrows S + NADH + H^+$

and is reversible, as shown. The net result of the forward reaction is the oxidation of the substrate, the reduction of the coenzyme, and the liberation of a hydrogen ion.

The biologic importance of substrate dehydrogenations is that energy has been transferred from the substrate to the reduced coenzyme. The oxidation of NADH by O_2 to restore NAD^+:

Equation 2–6 $NADH + \frac{1}{2} O_2 + H^+ \rightarrow NAD^+ + H_2O$

is accompanied by a $\Delta G^\circ = -52$ kcal per mole. In effect, the reactivity of the reduced dihydropyridine molecule represents about 35 to 50 per cent of the free energy change involved in the complete oxidation of a carbon atom of the substrate.

FLAVINS

The second coenzyme for enzymes catalyzing substrate dehydrogenations is flavin-adenine dinucleotide (FAD) (Fig. 2–1). The enzymes that FAD combines with dehydrogenate —CH_2—CH_2— groups, and transfer H atoms stereospecifically to FAD, in a manner analogous to the transfers with the pyridine nucleotides. The flavin-dependent apoenzymes bind the coenzyme quite strongly (K is of the order of 10^8 to 10^9 M), so that the flavin coenzymes do not apparently exchange readily between their several apoenzymes.

Succinate is oxidized to fumarate by a flavin-dependent dehydrogenase (Fig. 2–4). The succinate dehydrogenase is found bound to the membranes of mitochondria, where are also found flavoproteins that dehydrogenate fatty acyl-coenzyme A complexes, sarcosine, choline, and α-glycerophosphate. These mitochondrial enzymes are the most important in energy transformations; they do not react directly with O_2 but transfer free energy from substrate to $FADH_2$. There are other flavoprotein enzymes that are soluble and are found in the cytosol; they usually react directly with O_2 but are not of quantitative importance in energy transfer.

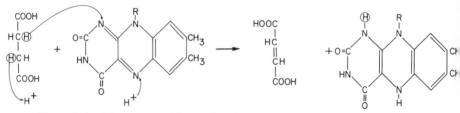

Figure 2–4.   The stereospecific transfer of an H atom, catalyzed by the FAD-dependent succinate dehydrogenase.

Mitochondrial FAD-dependent dehydrogenases transfer part of the free energy of oxidation of their substrates to $FADH_2$. Other mitochondrial electron-carriers, as will be discussed in the next chapter, oxidize $FADH_2$ through their eventual combination with O_2; the overall oxidation

Equation 2–7 $FADH_2 + \frac{1}{2} O_2 \rightarrow FAD + H_2O$

has a $\Delta G° = -37.3$ kcal per mole. The free energy liberated by the —CH_2—$CH_2 \rightarrow$ —$CH{=}CH$— reaction is transferred to an $={}N$—H bond of the isoalloxazine moiety of FAD. The negative free energy change of oxidation of $FADH_2$ is smaller, it will be noted, than that of NADH (-52 kcal per mole); that difference is important in electron-transport (See Chap. 3).

The enzymatic dehydrogenation mechanisms illustrate a general pattern for the transfer of energy in metabolic processes. The transfer is made through the use of intermediates common to two reactions; in this case (but not in all) the important intermediate is an enzyme molecule. Through such linkages, one reaction can drive another, if the first reaction is accompanied by a sufficiently negative change in free energy to meet the energy requirements of the second reaction. This principle will be better understood if the details of some enzymatically catalyzed dehydrogenations that are linked with the phosphorylation of ADP are examined.

SUBSTRATE-LINKED PHOSPHORYLATIONS

The enzymatic dehydrogenation of 3-phosphoglyceraldehyde in the presence of ADP and P_i to form ATP and 1,3-diphosphoglyceric acid is a prototype of reactions that partially transform the free energy of oxidation of substrates into the energy-rich $\sim$P-bond. Such reactions are called substrate-linked phosphorylations because they involve oxidations by substrate-specific enzymes. Other oxidative phosphorylations involve the oxidation not of substrates but of the reduced coenzymes that are produced by substrate oxidations; this type of oxidative phosphorylation takes place in membrane-bound systems. Substrate phosphorylations occur in solution. Because the understanding of chemical events in solution is much more advanced than that of chemical events on surfaces or in the crystalline or solid state, the dehydrogenations in solution are best understood at present. Actually, as will be seen, dehydrogenations in the semi-solid state account for the major portion of the cell's energy transformations. The soluble dehydrogenases serve as models for what we think—or hope—happens in the insoluble systems that catalyze electron-transfer (see Chap. 3).

The oxidation of 3-phosphoglyceraldehyde is catalyzed by the NAD^+-dependent 3-phosphoglyceraldehyde dehydrogenase (also called glyceraldehyde 3-phosphate dehydrogenase and triose phosphate dehydrogenase). The reaction the enzyme catalyzes has historical importance as one of the two enzymatic reactions studied by Warburg and Christian (1936) that established the specific functions of the pyridine nucleotide coenzymes in dehydrogenations and in energy-transfer (the other was the "Zwischenferment," the $NADP^+$-dependent glucose-6-phosphate dehydrogenase), and as an early demonstration of a redox reaction that is "coupled" with the synthesis of ATP from ADP and P_i. The reaction is one of the two oxidations of the anaerobic glycolysis pathway that transforms energy to $\sim$P-bonds.

The reaction catalyzed by phosphoglyceraldehyde dehydrogenase is:

Equation 2–8

$$
\begin{array}{l}
\text{CHO} \\
| \\
\text{HCOH} \\
| \\
\text{H}_2\text{C} \quad \text{O}^- \\
\qquad | \\
\text{O—P=O} \\
\qquad | \\
\text{OH}
\end{array}
\; + \text{NAD}^+ + \text{HPO}_4^{2-} \rightleftharpoons \;
\begin{array}{l}
\qquad\quad \text{O}^- \\
\qquad\quad | \\
\text{O=C—O}\sim\text{P=O} \\
| \qquad\quad | \\
\text{HCOH} \quad \text{O}^- \\
| \\
\text{H}_2\text{C} \quad \text{O}^- \\
\qquad | \\
\text{O—P=O} \\
\qquad | \\
\text{OH}
\end{array}
\; + \text{NADH} + \text{H}^+
$$

D-glyceraldehyde-
3-phosphate

1,3-diphospho-
glycerate

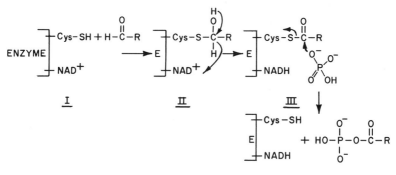

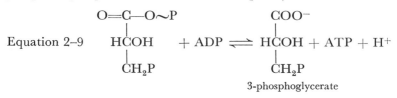

Figure 2-5. Steps in the oxidative phosphorylation of glyceraldehyde-3-phosphate to 1,3-diphosphoglycerate, catalyzed by 3-phosphoglyceraldehyde dehydrogenase. (After Jencks, 1963.)

A product of the oxidation is 1,3-diphosphoglycerate, an acid anhydride that contains an energy-rich phosphate group (see Fig. 1–1). The energy for the formation of this bond is derived from the negative free energy change that occurs when an aldehyde is oxidized to a carboxylic acid.

The complex reaction catalyzed by 3-phosphoglyceraldehyde dehydrogenase proceeds in steps. The final details of the mechanism are not yet sure, but a well-supported scheme is shown in Figure 2–5.

The enzyme contains in its peptide chain a cysteine, the sulfur of which binds the carbonyl group (I) to form a hemithioacetal (II). This complex is oxidized, two electrons being removed and a hydrogen atom being transferred directly to the enzyme-bound NAD^+, to form NADH (III). The free energy of oxidation of a C—H bond of the aldehyde has been transferred to a new high-energy compound, a thiol ester. The energy-rich thiol ester now reacts with inorganic phosphate and a free acyl phosphate, 1,3-diphosphoglycerate, is formed ($\Delta G° = -11.8$ kcal per mole).

In the cell, the ∼P-bond of 1,3-diphosphoglycerate can be transferred to ADP to form ATP through the agency of phosphoglycerate kinase (the phosphate group is summarized as P in Eq. 2–9).

$$
\begin{array}{c}
O{=}C{-}O{\sim}P \\
| \\
Equation\ 2{-}9 \qquad HCOH \qquad + ADP \rightleftharpoons HCOH + ATP + H^+ \\
| \\
CH_2P
\end{array}
\qquad
\begin{array}{c}
COO^- \\
| \\
\\
| \\
CH_2P
\end{array}
$$

3-phosphoglycerate

This is a thermodynamically feasible reaction, since the phosphorylation of ADP involves a free energy change of about 7 kcal per mole, whereas the dephosphorylation of the diphosphoglycerate liberates almost 12 kcal per mole. The *phosphokinase* enzymes, in general, catalyze such transfer reactions that require no additional energy input or are energetically favored, and they transfer ∼P-groups from various compounds to form the common intermediate ATP.

A second soluble enzyme reaction that involves the transformation of energy to an energy-rich bond is the formation of phosphoenol pyruvate from 2-phosphoglycerate, catalyzed by enolase. This is not a simple dehydrogenation and does not involve a coenzyme; it is a dehydration and internal oxidative dismutation that produces an enol phosphate, one of the class of high-energy phosphate bonds:

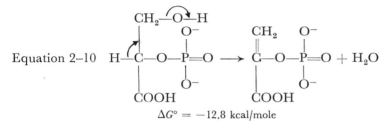

Equation 2–10

$$\Delta G° = -12.8 \text{ kcal/mole}$$

The energy-rich phosphate bond is formed through the conversion of the terminal carbon from an alcoholic to a ketonic carbon. Although Lehninger and Wadkins (1962) and Racker (1961) consider this reaction to be a type of substrate-linked oxidative phosphorylation and of significance because it is the only good instance of an energy-rich phosphate being formed through attachment of the phosphate *before* electron transfer occurs, Slater (1966a) thinks enolase has little relevance for the mechanism of oxidative phosphorylation because dehydration to form a C=C bond is not a typical redox reaction.

Other substrate-level phosphorylations involve the oxidative decarboxylation of pyruvate or α-ketoglutarate, and proceed by similar mechanisms that depend on the coenzyme thiamine pyrophosphate (Fig. 2–6).

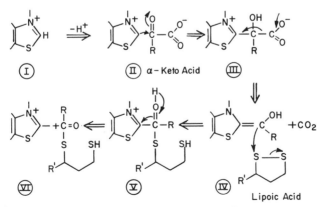

Figure 2–6. Steps in the oxidative phosphorylation of pyruvate or α-ketoglutarate, catalyzed by the pyruvate dehydrogenase or the α-ketoglutarate dehydrogenase complexes. (After Jencks, 1963.)

The thiazole ring of thiamine ionizes (I) and adds to the carboxyl group of the α-keto acid (II). The resulting unstable group rearranges (III) and undergoes decarboxylation (IV); the product attaches to enzyme-bound lipoate through an oxidation and forms an energy-rich thiol ester (V) which dissociates and frees the coenzyme (VI). These enzyme complexes are found in the mitochondrion.

Jencks (1963) points out that the substrate-linked reactions whereby oxidative steps generate energy-rich bonds are really of minor value to the cell's energy balance. For one thing, most of them are involved in the pathways of fermentation and glycolysis, which are minor sources of ATP for the mammal. As shown in Chapter 1, glucose oxidation by O_2 involves a free energy change of -686 kcal per mole; glucose yields only -56 kcal per mole via glycolysis.

General References: The pyridine nucleotide-dependent dehydrogenases are reviewed from different points of view: Colowick *et al.* (1966) and Strittmatter (1966) discuss various aspects of their enzymatic activities and mechanisms. Vallee and Hoch (1961) are concerned with the role of metals in their action. The flavoproteins are discussed by Massey and Veeger (1963). Jencks (1963) analyzes the chemical mechanisms of high-energy bond formation.

CHAPTER

3

ELECTRON-
TRANSPORT

The substrate-linked oxidative steps discussed in Chapter 2 involve the liberation and transformation of up to 30 per cent of the free energy of oxidation of a glucose molecule (see Fig. 1–2, Phase II). Substrate dehydrogenations lead to the formation of a few P-bonds and the reduced coenzymes NADH and $FADH_2$. The major portion of the negative change in the free energy of oxidation of substrate carbon atoms is involved in the metabolism of these two reduced coenzymes. Both NADH and $FADH_2$ are reoxidized in the mitochondrion, and the reducing equivalents so evolved pass along the steps of a common pathway, the electron-transport system, and finally reduce O_2 to H_2O. The steps whereby the electrons are transferred are thermodynamically favored, each stage being at a higher electron-transfer potential than its successor.

The free energy change between the two components of a redox system is measured accurately through the flow of electrons. Electron-transfer potential, another example of the group-transfer potentials mentioned in Chapter 1, is called *redox potential*. When two components of a redox system transfer electrons between each other, the redox potential is defined as follows:

Equation 3–1
$$\varepsilon° = \frac{\Delta G°}{\eta F}$$

where $\varepsilon°$ = the redox potential in volts; η = the number of electrons transferred in moles; and F = the caloric equivalent of the faraday, 23.06 kcal per mole of electrons. The redox potential is proportional to the change in standard free energy in redox reactions; at 25°, pH 7, a standard redox

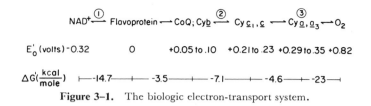

Figure 3–1. The biologic electron-transport system.

potential of 0.1 v is equal to $\Delta G° = 4.6$ kcal for 2-electron transfers. The electron-transport system is a series of pairs of electron-carriers with successively more positive redox potentials, i.e., with successively less negative changes in free energy.

As a guide to the ensuing discussion, Figure 3–1 shows the steps in the redox system, the free energy differences and the redox potentials. Between the initial oxidation of NADH and the final reduction of oxygen to water, there are three steps with a redox potential that represents a $\Delta G°$ between -7.1 kcal per mole and -23 kcal per mole. At each of those three steps the change in free energy is sufficient to support the formation of a high-energy bond; the rest of the free energy change produces heat.

The purposes of the biologic electron-transport system are *to transform energy to certain high-energy bonds and heat.*

Here we will discuss each type of redox component from the molecular point of view; in Chapter 4 it will become clear that these molecules are incorporated in a structural matrix in the mitochondrion that modifies their moleçular properties and imparts certain functional characteristics to the multimolecular complex.

The first step of electron-transport, after substrate-dehydrogenase action has produced NADH, is the transfer of electrons from NADH to a flavoprotein electron-carrier. FAD enzymes oxidize NADH:

Equation 3–2

$$NADH + FAD + H^+ \rightarrow NAD^+ + FADH_2; \qquad \Delta G' = -14.7 \text{ kcal/mole}$$

The relatively large $\Delta G'$ of this transfer reaction can be measured directly, through the concentrations of the reactants; it can also be calculated from the difference between the changes in free energy when NADH is oxidized by oxygen (-52 kcal per mole; Eq. 2–6) and when $FADH_2$ is oxidized (-37.3 kcal per mole; Eq. 2–7).

There are substrate-dehydrogenases that reduce FAD, the mitochondrial succinate dehydrogenase being the most active. The electrons of the $FADH_2$ so produced are transferred in the electron-transport chain to another flavoprotein carrier of equal or slightly more positive redox potential, and subsequently enter the same pathway as the electrons originating from NADH.

NONHEME IRON

Some substrate-specific dehydrogenases and some electron-transport carriers contain iron that is not incorporated into a porphyrin prosthetic group, that is, nonheme iron, in contrast to the cytochromes that contain prophyrin-bound heme iron. The identification of these enzymes has rested mainly on analytical methods for iron, and on the techniques of electron paramagnetic resonance (EPR) spectroscopy. EPR spectroscopy is based on the magnetic properties of electrons and can identify free radicals and paramagnetic metal ions and their interactions with their immediate environments. In practice, enzymes in mitochondria or purified enzymes have been placed in a magnetic field at low temperatures, and the amount of absorption of energy at different magnetic field strengths has been measured. A factor "g," the "spectroscopic splitting" factor, has been derived from the data which are determined by the intrinsic properties of the material under study.

In the first derivative of the energy-absorption spectrum, a striking signal is observed at $g = 1.94$, when whole fresh rat liver, liver mitochondria, submitochondrial particles, relatively purified NADH dehydrogenase, or succinate dehydrogenase is examined. The strength of the $g = 1.94$ signal depends upon the redox state of the enzymes; it is maximal at complete reduction, disappears at complete oxidation, responds to specific redox inhibitors, and has characteristics that make it unlikely that the signal could be due to an organic free radical. The evidence indicates that an iron-sulfur complex produces this signal. Not all nonheme iron produces the $g = 1.94$ signal, but these studies do show that some nonheme iron is involved in the main line of electron-transport and in some substrate-specific dehydrogenases. Several of the enzymes which exhibit the $g = 1.94$ signal are metalloflavoproteins, such as succinate dehydrogenase, xanthine oxidase from milk, aldehyde oxidase from rabbit liver, and dihydroorotic dehydrogenase from *Zymobacterium oroticum*. The nature of the iron-binding in the nonheme iron electron-carriers remains to be clarified. Apparently sulfur is involved; in acid solutions, these carriers release their Fe atoms and an equal number of sulfide ions. How this "labile sulfur" contributes to iron-binding in the carrier is not known.

UBIQUINONES: COENZYMES Q

The ubiquinones, or coenzymes Q, comprise a group of lipid-soluble benzoquinones that are found in most aerobic organisms, from bacteria to higher plants, and in mammals. The ubiquinones have a 2,3-dimethoxy-5-methyl benzoquinone nucleus, with a terpenoid side chain that contains one to 10 monounsaturated isoprenoid units. They are described either as ubiquinone X, where X is the number of carbon atoms in the side chain and

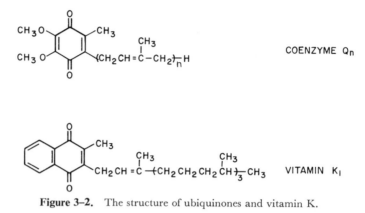

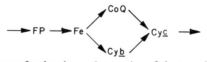

Figure 3–2.   The structure of ubiquinones and vitamin K.

can be any multiple of five up to 50; or as coenzyme Q_n, where n is one to ten depending on the number of isoprenoid units. The naturally occurring members are the coenzymes Q6 to Q10. Their structure may be compared with that of the naphthoquinone vitamin K, an antihemorrhagic factor (Fig. 3–2). The oxidized forms are shown in Figure 3–2; the reduced forms are hydroquinones.

At least one quinone is found associated with electron-transport systems of aerobic organisms; the type of quinone varies between different organisms, but in each, one type predominates. In mammals, both ubiquinones and forms of vitamin K are in the mitochondrion; in rats, ubiquinone also occurs in the other particulate fractions (nuclei and microsomes), but about two-thirds is in the mitochondrion (Aithal *et al.*, 1968). Mammalian mitochondria contain 6 to 8 molecules of ubiquinone per molecule of cytochrome *a*.

The exact role of the ubiquinones and vitamin K in electron-transport is still controversial, but it does seem clear that they participate either along the main route in the region between the flavodehydrogenases and cyto-chrome *b*, or as an equal alternative route with cytochrome *b* (Fig. 3–3). A difficulty in resolving between such possibilities has been that the measured rates of electron-transport via ubiquinones have not been fast enough to accommodate the overall observed rate of respiration. Thus, if ubiquinone is on the main chain, it would not permit rapid electron-transport, unless it is postulated that a very small fraction of the ubiquinone turns over electrons extremely rapidly but does not contribute much to the changes in the optical absorbance of all the ubiquinones in the mitochondria (the redox state of

Figure 3–3.   A scheme for the alternative routing of electrons between the flavoproteins and cytochrome *c* via either ubiquinone or cytochrome *b*.

ubiquinone is measured by the difference in absorption between the oxidized and reduced forms, like most of the other components of the cytochrome-containing respiratory chain). Klingenberg and Kröger (1967) present evidence that ubiquinone is on the main electron path, whereas cytochrome b is not, and stress the analogies between the roles of ubiquinone and NAD^+. Both coenzymes are relatively loosely bound and so are diffusible, but ubiquinone, being lipid-soluble, is confined to the mitochondrial membrane. Both NAD^+ and ubiquinone, being present in stoichiometric excess over the other components of the chain, function as H-collecting pools in the respiratory chain, with ubiquinone serving to collect from the flavodehydrogenases and NAD^+ from the pyridine nucleotide-dependent dehydrogenases. Chance *et al.* (1968a) review evidence that the function of ubiquinone can be accurately observed only in intact mitochondria, and that there the essentialness of ubiquinone remains to be proved.

Although it is not clear how ubiquinone participates in electron-transport, its participation appears to involve a primary site for the transformation of the electron-transfer potential. Ubiquinone is at or near a site for energy-coupling (see Crane and Löw, 1966, for a discussion of the evidence). There is controversy over whether vitamin K is also involved in energy-coupling, based on conflicting findings in chicks made dietarily deficient.

CYTOCHROMES

The cytochromes are, as their name implies, colored electron-carriers. They differ in some respects from the preceding components of the electron-transport chain. The cytochromes share the feature of having a prosthetic heme group as an active site (Fig. 3–4); that group is extremely firmly

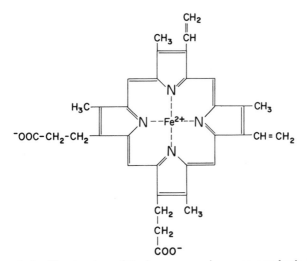

Figure 3–4. The structure of the heme group, iron protoporphyrin IX.

(covalently) bound to the apoenzyme proteins, and the combination is thought of as an entity, in contrast to the dissociable coenzymes of the components discussed up to now.

The hemoproteins are found in almost all phyla, and participate in respiration by transferring electrons (cytochromes), by carrying molecular oxygen (hemoglobin, myoglobin, erythrocruorin), or by reducing peroxides (in catalase and the peroxidases). The cytochromes of class b, the oxygen-carriers, catalase and peroxidases all contain the same heme: ferroprotoporphyrin IX, which is distinguished by the nature of the side-chains on the pyrrole rings. Cytochromes c_1 and c contain a heme that differs from ferroprotoporphyrin IX, and from the heme of cytochromes a and a_3. Not only do the hemes of the cytochromes vary in composition, but the protein moieties vary in their specificity; as a result, the Fe atom of each of these electron-carriers has a different redox potential and the heme has a characteristic absorption spectrum.

One of the characteristics used to identify the cytochromes is the position and intensity of the absorption maxima of the reduced pigment in the visible range. These peaks are contributed by the resonating rings of the porphyrin moieties, modified by the iron atom coordinated in their centers; by the constituent side chains on the pyrrole rings; and by the binding groups contributed by the protein apoenzyme. The spectrum modification contributed by the iron depends on the redox state of the iron, and so the absorption maxima shift when the iron changes between $Fe^{2+} \rightleftharpoons Fe^{3+}$.

The electron-acceptor and the electron-donor of the cytochromes is the coordinately bound Fe atom of the heme group. In the hemes, the four pyrrole nitrogen atoms are arranged in a plane and form a square-planar chelate complex with Fe. Fe has six coordination positions, and two are perpendicular to the plane of the porphyrin; one or both of these two positions are bound to a histidine or a methionine side-chain of the protein polypeptide chain, or to an —OH group.

The Fe atom of the heme exchanges one electron at a time: $Fe^{2+} \rightleftharpoons Fe^{3+} + e^-$. This imparts another distinctive difference between the cytochromes and the preceding electron-carriers (except, of course, the non-heme Fe components): the pyridine nucleotides, the flavoproteins, and the ubiquinones accept *two* electrons and an H atom from their donors and donate two electrons at a time to their acceptors. The problem of how a 2-electron transfer reduces a 1-electron acceptor in the respiratory chain is still not settled; it may involve the formation of cytochrome dimers or higher polymers, which can accept pairs of electrons, or the formation of flavin free-radicals which can donate one electron.

Through measurements of the redox potentials of each of the cytochrome components of the electron-transport chain, and by the use of specific inhibitors, the position of each cytochrome in the chain has been determined. As shown in Figure 3–1, the series is in the order Cy $b \rightarrow$ Cy $c_1 \rightarrow$ Cy $c \rightarrow$ Cy $a \rightarrow$ Cy $a_3 \rightarrow O_2$. The standard free energy changes at neutral pH between the steps include one of -7.1 kcal per mole between

Cy b and Cy c_1c, another of -4.6 kcal per mole between Cy c_1c and Cy aa_3, and a very large one of -23 kcal per mole between Cy a_3 and oxygen. Since the production of an energy-rich bond requires about 7 kcal per mole, there are several electron-transfer potentials in the cytochrome region that could support the synthesis of a $\sim$P-bond. Experimental observations show that two $\sim$P-bonds are actually formed via cytochrome electron-transfers, one at the Cy b-Cy c_1 site, the other between Cy c and O_2.

The assumptions that underlie the calculation of the changes in free energy include estimates of molecular concentrations and activities that are deduced from extinction coefficients obtained on purified components. The cytochromes may be polymerized, and some are firmly bound to lipid membrane components in the mitochondrion and estimations of their "concentrations" may well be inaccurate to a rather large degree; the derived calculations of free energy changes and redox potentials are rather risky. It will be noted that the transfer of electrons between Cy c and O_2 via the cytochrome oxidase involves a change in standard free energy of about -28 kcal per mole. The oxidation of Cy c (Fe^{2+}) by Cy a (Fe^{3+}) is accompanied by a $\Delta G°$ of only -4.6 kcal per mole; although there is spectral evidence that this is a locus of energy-transfer, it is thermodynamically more likely that the oxidation of Cy a (Fe^{2+}) by O_2 supports the formation of a $\sim$P-bond. The complicated composition and kinetics of the complex IV have not yet permitted a suitable hypothesis for the energy-transfer site. However, it is known that only one energy-transfer of about -7 kcal per mole occurs. The cytochrome oxidase is thus a major site of heat production, since about 20 kcal per mole of free energy change is not converted into another chemical form.

ELECTRON-TRANSPORT COMPLEXES

Relatively pure lipoproteins, of molecular weights between 250,000 and 400,000, can be prepared from intact electron-transport systems that function as if they were segments of the system and contain a small number of components. They are called Complex I, II, and so forth, or Site I, II, and so forth.

Complex I: Functions as an NADH:CoQ oxido reductase,

$$NADH + CoQ + H^+ \rightarrow NAD^+ + CoQH_2,$$

and contains NADH dehydrogenase (a flavoprotein), CoQ, and eight nonheme iron atoms.

Complex II: Functions as a succinate:CoQ oxido reductase,

$$Succinate + CoQ \rightarrow fumarate + CoQH_2;$$

it contains succinate dehydrogenase, eight nonheme iron atoms, and Cy b.

Complex III: A $CoQH_2$:Cytochrome c oxido reductase,

$$CoQH_2 + 2\ Cy\ c\ (Fe^{3+}) \rightarrow CoQ + 2\ Cy\ c\ (Fe^{2+});$$

it contains 2 Cy b, two nonheme iron atoms, and 2 Cy c_1.

Complex IV: A Cytochrome c:O_2 oxido reductase,

$$4\ Cy\ c\ (Fe^{2+}) + O_2 \rightarrow 4\ Cy\ c\ (Fe^{3+}) + 2\ H_2O,$$

containing cytochromes a and a_3, and Cu atoms; Complex IV is sometimes referred to as cytochrome oxidase.

These complexes should not be confused with the three or four sites of the electron-transport system that are involved in the transformation of energy into energy-rich bonds.

INHIBITORS OF ELECTRON-TRANSPORT

A variety of chemical agents inhibit electron-transport *in vitro* and poison respiration *in vivo*. Many of these act by binding to specific components of the electron-transport chain, and their specificity has been used to study the contribution of electron-carriers to respiration.

The categorization of the inhibitors of electron-transport is difficult for several reasons. Some of the inhibitors act at sites that are as yet incompletely characterized. With certain inhibitors, low concentrations affect one locus, and higher concentrations another locus as well; their specificity is not absolute. Inhibitory actions may depend not only upon the components of the electron-transport system, but also upon the degree of intactness of the system or upon the status of the phosphorylating processes associated with the system. Nevertheless, an approach toward classification can be made. Many of the inhibitors seem to act on one of the electron-transport complexes, and the site of action of some can be even further resolved (Fig. 3–5).

The "classic" respiratory poisons, cyanide (CN^-), carbon monoxide (CO), sulfide (S^{2-}), and azide (N_3^-) share the features of being small molecules that can form strong complexes with metal ions. They act on Complex IV, combining with the Fe^{3+} of the cytochrome oxidase complex, to inhibit respiration (Fig. 3–5, G). They can also form inhibitory complexes with a nonheme Fe atom, or a Cu atom associated with cytochrome a_3, or one of the other transition metal ions important in the actions of the metalloflavoproteins or the metallodehydrogenases. The affinity for Fe is so great that they also complex with hemoglobin and displace O_2 when in sufficiently high concentrations, and thereby interfere with oxygen-transport to the cell. *In vitro*, CN^- inhibits respiration at concentrations below about 10^{-4} M.

Several organic molecules inhibit reactions that occur in Complex I, between the flavoprotein carriers and coenzyme Q or cytochrome b or both

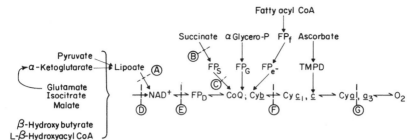

Figure 3–5. Hypothetical sites for the actions of agents that inhibit electron-transport. Abbreviations: FP = flavoprotein; Cy = cytochrome; CoQ = coenzyme Q; CoA = coenzyme A; TMPD = tetramethyl-p-phenylenediamine. Representative inhibitors are:
  (A) Arsenite; Cd^{2+}.
  (B) Malonate; oxaloacetate.
  (C) Thenoyl trifluoroacetone.
  (D) Uncoupling agents (high concentrations).
  (E) Amytal; rotenone.
  (F) Antimycin A; BAL + O_2; 2-alkyl-4 hydroxyquinoline-N-oxide; 3(2-methyloctyl)-. naphthoquinone.
  (G) CN^-; N_3^-; S^{2-}; CO.

(Fig. 3–5, E). Among them are the *barbiturates*, especially amobarbital (amytal), which act *in vitro* at concentrations above about 10^{-3} M; and *rotenone*, a multicyclic molecule isolated from certain plant roots and used as a fish poison, which is extremely potent and inhibits in molar amounts equal to those of a specific flavoprotein in the chain (i.e., stoichiometrically). The protein *toxins* of *Pasteurella pestis* inhibit electron-transport in Complex I, by combining not with a carrier but with the structural protein; they apparently have a high affinity for heart muscle mitochondria, where they rapidly depress the electron-transport-dependent uptake of Ca^{2+} and P_i ions, and the resultant changes in myocardial contraction (which are reflected in the electrocardiogram) are lethal (Kadis *et al.*, 1969).

Thenoyl trifluoroacetone, at concentrations about 10^{-4} M, inhibits processes in Complex II, and prevents electron-transport arising from succinate oxidation (Fig. 3–5, C).

A varied group of agents acts on Complex III, between cytochrome b and cytochrome c_1 (Fig. 3–5, F). The most potent are the *antimycins* (antimycin *a* is usually used), toxic antibiotics obtained from *Streptomyces*, that are effective in amounts stoichiometric with the cytochromes. The 2-alkyl-4-hydroxyquinoline-N-oxide inhibits at about 10^{-6} M, 3(2-methyloctyl) naphthoquinone at 10^{-5} M (but perhaps not at precisely the same site). The chelating agent "British anti-Lewisite" (BAL; 2,3-dimercaptopropanol) acts around this region when O_2 is present, at about 10^{-3} M.

Arsenite, trivalent arsenicals, or Cd^{2+} ions inhibit the dihydrolipoyl dehydrogenase enzyme complex by forming stable chelates with pairs of free sulfhydryl groups that are necessary for catalysis. These agents are thus specific inhibitors for the dehydrogenation of pyruvate and α-ketoglutarate, and the connected respiration (Fig. 3–5, A). Competitive inhibitors of substrate dehydrogenases, like oxaloacetate or malonate for succinate dehydrogenase, fall into an analogous category (Fig. 3.5, B).

Oligomycin, atractylate, aurovertin, valinomycin, guanidine, triethyltin, and HN_3 inhibit electron-transport arising from the oxidation of pyridine nucleotide-dependent substrates or succinate. They are not usually thought of as inhibitors of electron-transport, however, because they act only upon phosphorylating electron-transport systems (see Chap. 5).

Uncoupling agents in high concentrations inhibit electron-transport, although at lower concentrations they inhibit only phosphorylation and actually increase the rate of electron-transport (see Chap. 5). Their inhibitory actions on respiration are apparently connected with decreases in the transport of substrates into mitochondria and losses of intramitochondrial substrates, and thus represent a deficiency of electron-sources for the transport system (see Fig. 3–5, D).

Anaerobiosis inhibits electron-transport, because O_2 is the terminal electron-acceptor.

General References: The components of the electron-transport chain are described in Lehninger's books (1964, 1965), by Green and Baum (1970), by Ernster, Lee, and Janda (1967). For experimental approaches, see Chance and Hollunger (1963a, b, c) and Chance, Williams, and Hollunger (1963).

CHAPTER

4

THE
MITOCHONDRION

The *mitochondrion* is an organelle found in all mammalian cells. The name derives from the Greek *mitos*, thread, and *chondros*, grain, and it was used to describe the granular or filamentous particles seen with the light microscope by the histologists of the latter half of the nineteenth century. During the first half of the twentieth century, it became clear that the mitochondrion was functionally an oxidizing and energy-transducing apparatus that supplied energy in a utilizable form to cytoplasmic, nuclear, and ribosomal processes. The mitochondrion is a highly structured and complex organelle, which contains an almost complete set of biologic equipment: limiting membranes, information-storing molecules, insoluble Ca-P salts that resemble those in bone, the ability to change volume and to synthesize proteins, and a respiratory apparatus. Indeed, such completeness reminds one of the preformation theory of the biologists of the 1700's like Harvey, Malpighi, Bonnet, and von Haller, who proposed that life arose from the activation of a microscopic but complete precursor of an animal, which was presumed to reside in the head of the sperm. The mitochondrion resembles the homunculus. That such a resemblance is more than superficial will become more clear when the origin of mitochondria is considered.

Like an entire intact organism, the mitochondrion can be discussed from the point of view of its composition, anatomy or structure, physiology or function, and regeneration. Some of the information can be based on observations of mitochondria *in situ* in living or fixed cells. Information is also derived from mitochondria isolated from ruptured cells and sedimented by centrifugation in an appropriate aqueous medium; the studies on isolated mitochondria are based on the assumption that the properties are not changed, and in many, but not all, cases this seems to be a well-founded assumption.

31

COMPOSITION

Analysis of isolated mitochondria shows a number of resolvable molecular components: (1) Enzymes specific for substrate molecules; (2) enzymes specific for electron-transport and phosphorylation; (3) diffusible or free coenzymes specific for the dehydrogenases and other enzymes; (4) inorganic ions specific for some of the enzymes, or present in other roles; (5) lipid-containing membranes; (6) a "structural" protein that can combine with other components to form a functioning complex; (7) information-storing molecules, RNA and DNA; some of these may be organized in supramolecular particles like ribosomes; and (8) substrates, metabolic intermediates, organic metabolic cofactors, ATP, ADP, AMP, and others.

The molecular properties of the enzymes that act as substrate dehydrogenases and electron-transporters are discussed in Chapters 2 and 3. In the mitochondrion, the enzymes are not all in a free diffusible form, but some are bound to membranes and to a structural protein. The same is true of the pyridine nucleotide and ubiquinone coenzymes. Their functional properties are thus modified.

The incorporation of the enzymes and coenzymes in a structural matrix is reflected as well by their relative stoichiometries. As would be expected in a functioning organelle, the amount of each component is in a definite proportion to each of the others. The components most easily measured are those that have cofactors with distinctive spectra; thus, there is quantitative information on the mitochondrial content of pyridine nucleotides, flavins, ubiquinones, and hemes. This information gives the amount of each coenzyme and, because the hemes are so firmly bound to their apoenzymes, the amount of the cytochromes. The protein apoenzymes of the mitochondrion have no distinctive spectral or other features, and so cannot be measured accurately, except by comparison of enzymatic activities in the purified state and in the mitochondrion. The enzymes in the phosphorylating pathways do not, to our knowledge, contain or bind absorbing cofactors, and their stoichiometry is still not known. Table 4–1 shows the amount of respiratory carriers in isolated rat liver mitochondria, measured per g of mitochondrial protein and per mole of cytochrome a. (Cytochrome a is used as a baseline for the reasons given above; in these and other mitochondria, cytochrome c_1 is sometimes used.)

In rat liver mitochondria—and in the mitochondria obtained from other organs of the rat, and from other species—the cytochrome enzymes have an approximate 1:1 molecular stoichiometry to each other. The cytochrome content per g of mitochondrial protein is, however, not constant among mitochondria from different sources, or even from the different organs of the same animal. The coenzymes, in contrast, are present in amounts greater than cytochrome a. On a teleological level, the excess of coenzyme is reasonable, because there are many coenzyme-requiring enzymes that act on a

Table 4-1 *The Amount of Electron-Transport Components in Rat Liver Mitochrondria* *

RESPIRATORY CARRIER	$\dfrac{\mu \text{ MOLES}}{\text{g PROTEIN}}$	$\dfrac{\mu \text{ MOLES}}{\mu \text{ MOLE Cy } a}$
NAD+	3.8	19
Flavoprotein	0.72	3.6
Ubiquinone	2.1	10.5
Cytochrome b	0.18	0.9
Cytochrome $c + c_1$	0.34	1.7
Cytochrome a	0.20	1
Cytochrome a_3	0.22	1.1

* From Chance and Hess, 1962; Aithal *et al.*, 1968.

variety of substrates, whereas there is only one final common pathway for electron-transport.

It appears to be a generality that the mitochondrial electron-carriers exist in constant low molecular proportions, and that the coenzymes are relatively more plentiful (although their proportions also vary among different tissues). That generality indicates some sort of supramolecular organization, much as the observations of constant low atomic proportions in chemistry indicated the existence of compounds. In mitochondria, the postulated organized electron-transport unit is called the *respiratory assembly*.

Small sedimentable particles can be isolated from mitochondria disrupted by mechanical or chemical means. These subparticles contain lipid membranes, a few substrate-specific enzymes, and electron-transport and phosphorylating components. In them, the proportions of the cytochromes are still 1:1 as in intact mitochondria, but some of the coenzymes and protein enzyme moieties not firmly bound to membranes have been lost.

Mitochondria contain DNA (Nass and Nass, 1963; Schatz *et al.*, 1964; Kalf, 1964). In studies on mammalian mitochondria, those prepared from lamb hearts contain about 4 μg of DNA per mg of protein, and those from rat liver about 2.4 μg per mg (Kalf, 1964). The function of this DNA and of the variable amounts of RNA is discussed on page 40.

The amount of inorganic elements in intact isolated mitochondria does not appear to be in the same constant proportions as the electron-carriers (Table 4-2). There are probably several reasons for this. Many of the cations like Na+, K+, Mg²⁺, Ca²⁺, and Mn²⁺ are relatively loosely bound by mitochondria and can be taken up or released depending upon the functional state of the mitochondria. Some of the cations of the transition metals are prosthetic groups for respiratory enzymes: Fe in the cytochromes and in the nonheme Fe-containing carriers, Zn in some of the pyridine nucleotide-dependent dehydrogenases, and Cu in a portion of the cytochrome oxidase (cytochrome a_3). As stoichiometric components of enzymes, this part of the mitochondrial metal content is constant; however, there probably is metabolic exchange,

Table 4–2 *Inorganic Elements in Rat Liver Mitochondria**

ELEMENT	μg ATOM g PROTEIN	μg ATOM μ MOLE Cy a
Na	59.8	299
K	107	535
Mg	26.5	133
Ca	4.21	21.1
Fe	2.35	11.8
Zn	0.34	1.7
Cu	0.16	0.8
Mn	0.11	0.6

* Recalculated from Thiers and Vallee, 1957; and Table 4–1.

for example, of free Fe^{2+} and Zn^{2+} ions as well, and that part of the metal content will vary.

COMPARTMENTATION

A description of the contents or compositions of mitochondria gives only a listing of the atoms and molecules that may be available to serve the function of mitochondria. Such a description is incomplete, especially for mitochondria, because the structure and the metabolic state of the mito- chondrion impose restrictions on the availability of the components to each other and to the outside environment and on the behavior of the com- ponents. The mitochondrion is not a bag containing a solution of mixed enzymes and ions but an organized entity. The membranes enclose a domain which imposes its own rules on the behavior of functioning molecules, and those rules are not the same as the ones which pertain outside the membrane. These differences have been ascribed to compartmentation or crypticity of the mitochondrial components, which is another way of describing the phenomenon itself but does not necessarily mean that there are physically separate compartments enclosed by limiting membranes.

An early demonstration of compartmentation was when Lehninger (1951) showed that isolated rat liver mitochondria could oxidize substrates that were initially dehydrogenated by pyridine nucleotide-dependent dehydrogenases but could not oxidize NADH itself. The endogenous intra- mitochondrial NAD^+ was reduced by such substrates and then the endog- enous NADH was oxidized by the electron-transport chain. The substrate had access to the dehydrogenase, but externally added NADH had no access to the "NADH oxidase." Studies with radioactive nucleotides have since shown that there is little if any exchange between endogenous and exog- enous NADH. That inaccessibility appears to involve the structure of the mitochondrion. The submitochondrial particles obtained from sonicated mitochondria oxidize NADH readily. The mechanical rupture of the membrane structure has altered the compartmentation properties.

It is perhaps not unexpected that the outer membranes of the mitochondrion should be selectively permeable and permit entry of substrates but not NADH. That is the sort of outside-inside compartmentation that any discrete organelle would exhibit; however, there is another form of compartmentation that is of greater significance for understanding mitochondrial function that might be called "inside-inside" compartmentation. One endogenous component does not seem to be available to another endogenous component. Such compartmentation would not be expected, unless there are specific internal structural or binding forces.

For instance, rat liver mitochondria contain both NAD^+ and $NADP^+$ in about equal amounts. They also contain a large amount of glutamate dehydrogenase. When that glutamate dehydrogenase is isolated and purified, it requires a coenzyme, but it has about equal affinity for NAD^+ and $NADP^+$ when it dehydrogenates and deaminates glutamate or acts in the reverse reaction. In intact mitochondria, however, glutamate dehydrogenase apparently reduces only $NADP^+$ when oxidizing glutamate or oxidizes only NADPH when acting in the reverse reaction (Borst and Slater, 1960; deHaan *et al.*, 1967). This compartmentation of the enzyme and the coenzymes might arise either from a spatial separation from NAD^+ inside the mitochondrion, or from an intramitochondrial binding of the enzyme that results in an altered specificity toward the two coenzymes. Whatever the explanation is, the phenomenon of compartmentation alters metabolic pathways and so alters energy metabolism.

Compartmentation has yet another aspect besides its control over energy metabolism. In some cases, compartmentation is controlled by the amount of available energy. Again, the glutamate dehydrogenase reaction is an example, but there are others as well. There is evidence that despite the large amount of glutamate dehydrogenase in rat liver mitochondria, the route of transformation of glutamate to α-ketoglutarate is not that catalyzed by glutamate dehydrogenase (GDH) (Eq. 4–1), but an entirely different one catalyzed by a transaminase (TA in Eq. 4–2).

Equation 4–1

$$\text{Glutamate} + \text{NADP}^+ \xrightarrow{\text{GDH}} \alpha\text{-Ketoglutarate} + \text{NH}_3 + \text{NADPH} + \text{H}^+$$

Equation 4–2

$$\text{Glutamate} + \text{Oxaloacetate} \xrightarrow{\text{TA}} \alpha\text{-Ketoglutarate} + \text{Aspartate}$$

That routing is dependent upon a source of energy—ATP. When uncouplers are added, the transamination pathway is inhibited; then the oxidative pathway (Eq. 4–1) becomes predominant. It is suggested that the compartmentation barriers are energy-dependent, and that oxaloacetate, which inhibits the malate and other dehydrogenases, plays an important role, because its active translocation from one part of the mitochondrion to another requires energy.

Again, whatever the mechanism, the internal compartmentation of mitochondria is controlled by the level of available energy. In turn, compartmentation controls the function of mitochondria in producing available energy. This sort of feedback control is one of the characteristics of biologic systems.

STRUCTURE

Mitochondria are discrete organelles of a size and shape that varies with the function of the tissue in which they are found, presumably a reflection of their specialization as part of the cellular apparatus. Mitochondria are usually asymmetric. Their diameter is between about 0.5 and 1μ. When examined in the cells of mammalian liver, kidney, or pancreas, they are about 3μ long, but the asymmetry is greater in the exocrine cells of the pancreas, where they are about $10\ \mu$ long.

The mitochondrion has a limiting surface membrane, composed of a lipid-protein bilayer and an inner membrane that is different in structure. The outer membrane has a lower density (Parsons, Williams, and Chance, 1966). The inner membrane projects into the matrix of the mitochondrion in many folds, the *cristae mitochondriales*. Projecting from the *cristae* and attached to them via a short stalk are small globular particles that are not seen in the outer membrane. The convolutions contributed by the *cristae* greatly add to the total surface area of the membrane. Lehninger calculates that the surface or outer layers of the mitochondria are about four times the surface area of the typical liver cell in which they are found, and that when the *cristae* are included, the mitochondrial membranes are about 10 times more extensive than the cell membrane. The surface of the stalked globules adds yet more area. Thus, the respiratory assemblies, which are attached to or are in the inner membrane, are spread over a relatively large surface, which modifies the properties of their constituent enzymes and the rates of diffusion or transport of fuels and products.

The inner and outer membranes of mitochondria from rat liver also differ in the ability to incorporate amino acids into their proteins *in vitro*. Only the inner membrane takes up ^{14}C-leucine. There is evidence that *in vivo* the inner and outer membranes are synthesized by different systems. The enzymic constitution of the outer membrane resembles in some ways microsomal membranes; the inner membrane is sensitive to the action, e.g., of chloramphenicol (like bacterial, but unlike microsomal, membranes) and is affected by cytoplasmic mutation. Roodyn and Wilkie (1968) suggest that the synthesis of the outer membrane may be under direct nuclear control, and that the synthesis of the inner membrane may be under the control of MDNA (mitochondrial DNA), as if the inner membrane had evolved from a primitive "bacterial" membrane which has been enclosed in an outer membrane derived from the host cell. Some workers consider the mitochondrion to be an "endosymbiont" vestige, remaining after an ancient invasion of evolving cells by a parasitic microbe (see p. 40).

There are about 2500 mitochondria per rat liver cell, and they normally make up about 20 per cent of the volume of the whole cell. As might be expected of an organelle with a semipermeable-limiting double membrane, the structure of the mitochondrion can vary with the composition of the external solution, with the metabolic state of the mitochondrion (i.e., with the composition of the internal phases), and under the influence of hormones and drugs.

One approach to understanding the mechanisms of oxidative phosphorylation has been to take apart the mitochondrion and attempt to put it back together. Attention has been focused on the inner membrane, since it contains the enzymes for electron-transport and phosphorylation. Fractionations of mitochondria after a mild disruption produce small particles that are still able to respire and phosphorylate. More drastic methods of disruption produce particles that contain the enzymes for respiration but not those for phosphorylation. The phosphorylating enzyme or cofactors have thus been solubilized or inhibited during the fractionation procedure; because it is easier to purify soluble materials than insoluble ones, such observations have enabled the isolation of new components and complexes from mitochondria that are involved in processes other than just electron-transport.

A number of substances have been isolated by fractionating mitochondria and purified to various degrees that have special effects on features of oxidative phosphorylation, and when combined can reconstitute the function of the intact mitochondrion. They have been called "coupling factors." Here the semantic usage of "coupling" refers to the "trivial" sense discussed in Chapter 5: the isolated factors couple phosphorylation to oxidation but do not necessarily make the rate of oxidation dependent on the phosphorylative metabolic state (i.e., couple in the sense of imparting respiratory control). The coupling factors have been identified as F_n, where n is a number to identify the order in which they were isolated.

A fraction composed of membrane fragments, which has no properties of a coupling factor, contains the entire respiratory chain, plus phospholipids and insoluble "structural" protein(s). When the cytochromes and phospholipids are removed from this insoluble fraction, it "masks" or suppresses the ATPase* action of a soluble coupling factor (F_1). Adding the phospholipids to the mixture restores the ATPase activity and makes the ATPase sensitive to inhibition by oligomycin.

---

* Adenosine triphosphatase; the term implies a protein with catalytic action, that accelerates the energy-wasting hydrolysis of ATP: $ATP + H_2O \rightarrow ADP + P_i$ (see Eqs. 1–7 and 5–1). The so-called "ATPase" may be a single molecule like the heavy meromyosin portion of myosin, or a multicomponent system that utilizes ATP. In both cases, ATP produces ADP and P_i eventually; energy has not been wasted, but rather transferred. The importance of a molecular "ATPase" is that in the reverse direction, the catalyst would synthesize ATP from $ADP + P_i$; a "reverse ATPase" plays a central role in several theories of oxidative phosphorylation.

F_1, as implied above, has ATPase activity. It has been purified to a considerable degree, and is a soluble protein of molecular weight about 280,000. F_1 loses its activity at $0°$, an unusual feature for an enzyme. This protein has an interesting relationship with oligomycin. Oligomycin (see p. 55) imparts a phosphorylative capacity to certain kinds of nonphosphorylating submitochondrial particles and in that context is itself a "coupling factor." Antibodies prepared against F_1 block the coupling-factor activity of oligomycin, suggesting a similarity between F_1 and oligomycin. That similarity is not complete, however, and F_1 acts differently from oligomycin in other systems.

Racker's laboratory has reported the resolution of mitochondrial proteins that stimulate either overall oxidative phosphorylation or specific partial reactions of oxidative phosphorylation. A "structural" protein has the property of conferring oligomycin sensitivity, i.e., an inhibitory action, to another protein fraction that is an ATPase; the structural protein is prepared in various states of polymerization. Yet another fraction is involved in either the formation or the stabilization of nonphosphorylated high-energy intermediates.

A number of semipurified preparations with properties partly similar to Racker's factors, as well as some with different properties, have been studied in various laboratories (Pullman and Schatz, 1967). Beyer (1963) has isolated ATP synthetases. A soluble ADP-ATP exchange enzyme has been purified that restores phosphorylation at Site III in salt-extracted mitochondria. It appears that the system catalyzing oxidative phosphorylation soon will be put together *in vitro* from soluble-resolved components.

An important concept has evolved from the observations on coupling factors, namely that the special properties of some of the enzymes in intact mitochondria can be reproduced through the interaction of purified soluble enzymes with membrane or insoluble fragments. The binding of soluble enzymes to a membrane usually alters the properties the enzyme shows in solution. Racker terms this process *allotopy*.

Mitochondria respond to changes in their aqueous environment by changing their volume. Mitochondrial swelling and contraction may be *passive*, in which the semi-permeable membrane imparts rapid and predictable responses to changes in osmolarity of the medium, or they may be *active*, in which respiration, metabolic state, and specific exogenous factors control changes in volume that may be slow or rapid. Either type of volume change is usually measured by the light scattering of suspensions of isolated mitochondria. The degree of light scattering depends upon the refractive index of the particles as compared with the medium. Electron micrographs of tissue sections demonstrate *in situ* swelling.

Active swelling is not seen in all mitochondria; for example, mitochondria from the brain show little if any swelling. In susceptible mitochondria, a variety of agents induces swelling: Ca^{2+} ions, phosphate ions, glutathione, ascorbate, the higher fatty acids (U factor), and thyroid hormones. Many of these agents are uncouplers (see p. 53), but not all uncouplers are swelling agents; 2,4-dinitrophenol, the classic uncoupler, produces no swelling. The swelling induced by the active agents is prevented or reversed by bovine serum albumin, some chelating agents, certain respiratory inhibitors, and ATP.

There are two different types of active swelling, according to the time required for maximal change and the degree of change in volume (Packer and Tappel, 1960). Low-amplitude swelling (Fig. 4–1) is rapidly responsive to respiratory changes, showing little or no lag period, cycling in

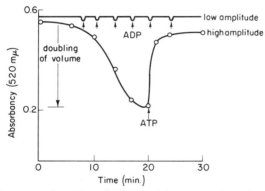

Figure 4–1. Two types of swelling of mitochondria, measured as changes in the apparent absorption of light at 520 mμ. (After Lehninger 1964; by permission of the author and W. A. Benjamin, Inc., New York.)

1 to 2 minutes, with a small amplitude (about 1 to 5 per cent of initial optical density). High-amplitude swelling (Fig. 4–1) is slow, with a lag period and maximal change up to 60 minutes, and an increase of 45 per cent in average mitochondrial diameters measured by electron microscopy which corresponds to a 100 to 200 per cent volume change for spherical mitochondria. Losses of respiratory control, phosphorylative efficiency, and the ordinarily tightly bound mitochondrial pyridine nucleotides accompany high-amplitude swelling.

Active swelling depends upon electron-transport and requires an electron source (a substrate) and an electron acceptor (O_2, or in a cyanide-poisoned system, ferricyanide, for instance). Electron-transport inhibitors like cyanide, amytal, antimycin *a*, or oligomycin can prevent the induction of swelling by the specific agents mentioned above. This dependence of active swelling upon the redox state of the carriers is taken as evidence for an "energy-linked" process.

The water that is taken up by mitochondria in high-amplitude swelling apparently goes to the soluble matrix inside the inner membranes. As the amount of water increases, the *cristae* that are so obvious in normal mitochondria disappear or are stretched out; fully swollen mitochondria have few or no *cristae.*

Low-amplitude swelling may not represent a volume change at all. The light scattering methods depend upon differences in refractive index between mitochondria and their suspending media; although high-amplitude changes in light scattering have been correlated with actual increases in water content of mitochondria, the low-amplitude changes are difficult to correlate with any microscopically visible change in volume of similar magnitude. Further, there is evidence that the observed changes in optical density might easily arise from changes in the internal structure of mitochondria, as shown by electron densities in electron microscopic studies (Hackenbrock, 1966). The relation between these changes and the metabolic states of mitochondria are discussed in the next chapter.

PROTEIN SYNTHESIS AND THE BIOGENESIS OF MITOCHONDRIA

Increasing interest in the origin and the mechanism of replication of mitochondria has been shown in the last few years. Mitochondria can synthesize proteins both *in vivo* and *in vitro.* They incorporate radioactive amino acids into proteins in the mitochondrion. The mitochondrion contains a set of protein-synthesizing equipment, which includes DNA, DNA-dependent RNA polymerase, tRNA, amino acid-activating enzymes, and ribosomes. This self-sufficiency suggests that the mitochondrion can replicate itself as well as synthesize its own proteins. Indeed, there are proposals that the mitochondrion is a descendant of a parasite that invaded cells long ago, subsisted symbiotically, and has since become an intrinsic contributor to the cellular ecology.

Part of the evidence for such an alien origin is that the DNA in the mitochondrion is not like mammalian nuclear DNA, as shown by analysis of its bases and their distribution, but more like bacterial or viral DNA. Mitochondrial DNA is a double-stranded ring, unlike mammalian DNA (Naas, 1966). And the characteristics of the mitochondrial incorporation of amino acids differ from those in the cytoplasm: mitochondrial protein synthesis does not require cell sap or pH 5 enzymes; is not inhibited by ribonuclease; is sensitive to a number of antibacterial antibiotics, particularly chloramphenicol, that do not inhibit cytoplasmic incorporation; and is insensitive to cycloheximide, an inhibitor of protein synthesis in mammalian endoplasmic reticulum ribosomes, but not in bacteria. It seems clear, therefore, that both the information-storing and the protein-synthesizing processes of mitochondria are qualitatively different from those of the rest of the mammalian cell but resemble those of bacteria.

Such findings have, in their turn, raised a different problem, and an upsetting one. Are the DNA and the rest of the information-storing and protein-synthesizing apparatus found in mitochondria merely reflections of bacterial contamination? Mitochondria are almost always isolated under nonsterile conditions and contain bacteria, up to 100,000 per ml in some reports. There is both direct and indirect evidence that the mitochondrion does contain its own synthetic equipment and does not owe that function to contaminating bacteria; it must be admitted, however, that some of the evidence against discarding a role for bacteria has not yet been controverted completely satisfactorily.

Making it unlikely that bacterial contamination contributes to mitochondrial protein synthesis are the observations that synthesis depends on the type of energy source supplied to the isolated mitochondria; succinate supports synthesis (it is a mitochondrial substrate) but glucose does not (and mitochondria alone cannot use glucose, athough bacteria can). The synthesis depends on added P_i, Mg^{2+} ions, and adenine nucleotides, as well as on the tonicity of the medium and the intactness of the mitochondria. There is no correlation between the rate of protein synthesis and the number of bacteria; and the incorporated amino acids sediment with the mitochondria, not the bacteria, in sucrose gradients. Injection of thyroid hormones stimulates amino acid incorporation (the author is not aware of any evidence that the hormone does not stimulate bacterial incorporation, however). Lastly, and perhaps most convincingly, germ-free mitochondria incorporate amino acids *in vitro* at reasonable rates.

It thus appears that bacterial contamination makes no significant contribution to mitochondrial protein synthesis; however, the observations of Sandell *et al.* (1967) and Beattie *et al.* (1967) must still be kept in mind. Mixtures of bacteria and inactive mitochondria show accelerated incorporation; the mitochondria apparently stimulate bacterial protein synthesis. Perhaps that stimulation is, as Roodyn (1968) suggests, merely a reflection of the bacteria being in a "poor metabolic state" and using mitochondrial cofactors to better themselves. On the other hand, there is evidence that mitochondria produce a water-soluble factor that accelerates ribosomal protein synthesis (Sokoloff *et al.*, 1968; see Chap. 8); mitochondrial stimulation of bacterial protein synthesis might be more specific than Roodyn suggests.

Even though mitochondria do synthesize proteins and even incorporate hexoses into glycoproteins *in vitro* (Bosmann and Martin, 1969), they do not replicate themselves completely. Mitochondrial DNA has a molecular weight too low to account for the coding of all the known mitochondrial proteins. Many of the functioning elements of the mitochondrion, like the cytochrome electron-carriers, are coded for and synthesized outside the mitochondrion presumably by the cell's cytoplasmic apparatus, and only then are they transported to the mitochondrion. The mitochondrion apparently does synthesize its own "structural" protein. It is possible, therefore, that the "structural" protein has open specific binding sites for the respiratory components, and that the cytochromes after entering the mitochondria literally find their proper places in the electron-transport complex. It is not too difficult to accept this concept at present, knowing what we do of the capacity of viruses, for example, to subvert the synthetic apparatus of host cells to synthesize viral components. This concept is consistent with the observed variable proportions of cytochromes and other respiratory components under the control of the thyroid hormones (see Chap. 8).

The evidence concerning the genetic controls over mitochondrial synthesis is derived mainly from experiments on yeasts and *Neurospora*, which are facultative anaerobes. There, cytochrome *c* is synthesized on the ribosomes in the cytoplasm under the direct control of genes in the nucleus. There is some evidence that cytochromes *a* and *b* are synthesized in mitochondria, presumably on the mitochondrial ribosomes under the influence of MDNA, but not all investigators agree. Whether the various cytochromes of mammalian mitochondria are synthesized at separate loci and then positioned in the proper sites on the mitochondrial structural protein is not yet clear, although cytochrome *c* does appear to originate outside the mitochondrion. It is striking that in the hormonal control of the cytochrome content of mammalian mitochondria, thyroid deficiency depresses the amount of *a*, *b*, and *c* cytochromes to about the same degree, and thyroid excess raises that amount to the same degree. Either thyroid hormone affects the rates of protein synthesis in mitochondria and in cytoplasmic ribosomes to just the same degree, or perhaps there is only one (extramitochondrial or intramitochondrial) site of cytochrome synthesis or control of synthesis. Thyroid hormones can act directly *in vitro* on mitochondria to stimulate their synthesis of proteins but not on ribosomes (cytoplasmic) or on nuclei. (Experiments measuring the rates of synthesis of each of the cytochromes in mitochondria from thyroxine-treated hypothyroid rats might resolve the problem; if the repletion of individual cytochromes proceeds at the same rate and stops when the same amounts are synthesized, it would indicate a single synthetic or control site.)

The importance of the biogenesis of mitochondria for understanding energy transformations is that the rate of biogenesis controls the absolute amount of respiratory apparatus in the cell (as does, of course, the relative rate of mitochondrial degradation). The rate of energy transformation depends upon two factors—the absolute amount and the specific respiratory activity of the respiratory apparatus. These two factors are subject to separate control mechanisms, and they can change independently or in a coordinated manner. Toxic agents change specific activity; hormones change both activity and amount purposefully.

General References: Several excellent reviews have recently appeared that discuss the mitochondrion from various points of view: Lehninger (1964), Racker (1965), Mitchell (1966, 1968), Green and Baum (1970).

Pullman and Schatz (1967) catalog and characterize a number of "coupling factors." Racker (1968) discusses his studies on the membrane of the mitochondrion and its recombining components. There is an increasing number of reviews on the origin and replication of mitochondria and their components: Roodyn and Wilkie (1968), Nass (1969).

CHAPTER

5

ENERGY
COUPLING

At three sites in the electron-transport chain, the redox potential differences are great enough—between -7 and -28 kcal per two moles of electrons transferred—to support the transformation of the free energy of reaction into the formation of an energy-rich bond, which in the $\sim$P-bond accounts for about -7 kcal per mole. The amount of energy-rich $\sim$P-bonds can be measured in a number of ways, and the experimental values are within the limits set by the theoretical considerations of thermodynamics.

In general, $\sim$P-bonds are measured in ATP, where they are formed through the esterification of inorganic phosphate ions (P_i) with the terminal (energy-rich) phosphate group of ADP. This phosphorylation reaction is:

Equation 5–1 $\qquad ADP + P_i^- + H^+ \rightleftarrows ATP + H_2O$

and its reverse, the hydrolysis of ATP that proceeds with a free energy change of -7 kcal per mole, is identical with Equation 1–7. The formation of $\sim$P-bonds is determined by measuring the incorporation of P_i into ATP chemically or through the use of radioactive ^{32}P. The newly esterified P_i is determined after separating ATP from the reaction mixture, or through the use of a P_i-trapping system like hexokinase-glucose-Mg^{2+}, which fixes the terminal $\sim$P of ATP into glucose-6-phosphate (where it does not react as P_i) and permits the estimation of new $\sim$P-bonds through the "disappearance" of P_i.

When mitochondria oxidize a substrate which donates its electrons and H atom at the level of the pyridine nucleotides in the transport chain (or when NADH is the substrate under special circumstances), three moles of $\sim$P-bonds are formed per mole of substrate oxidized. When succinate is the substrate, its electrons enter at the level of the flavin enzymes, and two moles of ∼P-bonds are formed per mole of succinate oxidized. Whatever the substrate, the oxidation of one mole involves the reaction of one-half mole of O_2 to form H_2O.

The ratio ∼P:O, usually written as P:O, is also called the *phosphorylation quotient*. Its units are moles of ∼P-bonds formed per gram atom of oxygen consumed. It is a measure of the efficiency of energy transformation in the sense of being the amount of utilizable energy produced per amount of energy consumed. The overall efficiency of mitochondria oxidizing a substrate that yields a P:O ratio of 3 can be calculated. If for each ∼P-bond, $\Delta G = -7$ kcal per mole, and for each $NADH + \frac{1}{2} O_2 + H^+ \rightarrow NAD^+ + H_2O$, $\Delta G = -52$ kcal per mole, the efficiency is $-21/-52 = 40$ per cent.

Thus 40 per cent of the change in free energy of oxidation of NADH has been transformed by the mitochondrion into ∼P-bonds. This portion of the free energy change is "utilizable" in the sense that ∼P-bonds of ATP can support energy-requiring reactions and processes in the mitochondrion and elsewhere. The other 60 per cent of the energy liberated is converted to heat or to nonphosphorylated energy-rich bonds; both are "utilizable," the energy-rich bonds for doing intramitochondrial work, and the heat, not for doing work, but for maintaining the body temperature of mammals, for instance, above that of their surroundings. "Utilizable" energy usually refers only to ∼P-bonds.

The anaerobic process of glycolysis has an efficiency of 27 per cent, with two ∼P-bonds being formed per mole of glucose broken down to lactate; however, the oxidation of glucose to CO_2 and H_2O involves a free energy change almost 20 times greater. The efficiencies of oxidation and glycolysis are not very dissimilar, but the total amount of utilizable energy produced is much greater when glucose is oxidized. This illustrates the point that efficiency is an index only of the relation between the input and output of energy but not of the net flux or the rate of transformation.

COUPLING BETWEEN RESPIRATION AND PHOSPHORYLATION

Biologic systems transform the energy of oxidation to a useful form. One measure of the cost of such oxidation is the amount of O_2 consumed. When only the *amount* of O_2 is considered, the performance of the biologic machine is described in thermodynamic terms as efficiency; however, efficiency does not involve the *rate* at which a machine operates. A hypothetical biologic system operating at a very low temperature, for instance, might consume 1 g atom of oxygen per year and convert the energy liberated to three moles of ∼P bonds. Its efficiency is the same as another system operating at 37°C and consuming 1 g atom of oxygen per hour. The capacity of those two systems to do work at any particular time is not measured by their thermodynamic efficiency but rather by the rate at which they produce useful energy.

It is therefore necessary to consider not only the amount of fuel a biologic system consumes, but also the rate at which it consumes the fuel and produces useful energy. Respiration is a general term that refers to the intake of oxygen. The whole mammal takes in oxygen at a rate determined by a number of factors, and transports that oxygen, bound to hemoglobin, to the membrane of cells, where it is liberated. Oxygen diffuses across the cytoplasm to the mitochondria, where it is utilized. We are concerned here with mitochondrial respiration, which accounts for more than 90 per cent of the total respiration of the intact mammal.

In the thermodynamic sense, respiration and phosphorylation in the mitochondrion are "coupled," or causally linked together. The First Law being what it is, some metabolic process that is sufficiently exergonic must support the synthesis of energy-rich phosphate bonds. In the mitochondrion, the free energy of oxidative reactions is used to produce ~P-bonds. When there is no oxidation of substrates, i.e., no transfer of electrons, no energy is transferred and no ~P-bonds can be generated. In general, the rate of mitochondrial respiration reflects the rate of oxidation, and *mitochondrial phosphorylation depends on respiration.* Lipmann (1941) used the term "coupling" in this sense in his now classic review on the metabolic generation of phosphate bond energy and pointed out that ~P-bonds can be generated via (extramitochondrial) anaerobic processes as well. This "coupling" of mitochondrial respiration and phosphorylation is, in a way, a trivial one.

However, this is not the only sense in which the coupling of respiration and phosphorylation is meant when considering mitochondrial metabolism. The converse is also true: The amount of ~P formed controls the amount and the rate of respiration. *The rate of respiration depends inversely on phosphorylation.*

This use of the term "coupling" is based upon experimental observations later than Lipmann's original concepts. When mitochondria oxidize substrates in the absence of ADP or P_i or both (or in the absence of a system that hydrolyzes the ATP formed to ADP + P_i), the rate of respiration is low (Fig. 5–1). That low rate of respiration is raised two- to 50-fold by adding

Figure 5–1. Respiratory control. Rat skeletal muscle mitochondria are added to a reaction mixture at pH 7.4 containing 0.28 M sucrose, 10 mM KCl, 4.5 mM P_i, 10 mM Tris, 0.22 mM ethylenediamine tetra-acetate. The ordinate is O_2-concentration, measured polarographically with a Clark electrode; the abscissa is time. The slopes of the recorded line represent the rate of respiration. The respiratory control index (State 3/State 4) is 3.55.

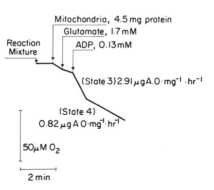

ADP + P_i; the degree of "respiratory control" is measured by the ratio of stimulated to resting respiratory rates; and when the added ADP + P_i is transformed to ATP, the stimulated respiration immediately reverts to the unstimulated slow rate.

The availability of ADP + P_i is not the only factor controlling the rate of mitochondrial respiration. Chance has analyzed a number of factors and stresses that the *"metabolic state"* of mitochondria, as influenced by ADP + P_i, oxygen, substrates, or the presence of inhibitors or uncouplers, controls the rate of respiration. He enumerates several such metabolic states in Table 5–1.

Table 5–1   *Some Metabolic States of Mitochondria**

STATE	O_2	SUBSTRATE	UNCOUPLER	$P_i + P^-$ ACCEPTOR	RESPIRATION
1	Excess	Endogenous	0	Endogenous	Slow
2	Excess	0	0	Excess	Very slow
3	Excess	Excess	0	Excess	Fast
3u	Excess	Excess	Excess	0	Fast
4	Excess	Excess	0	0	Slow
5	0	Excess	0	Excess	0
6	Excess	Excess	0	No P_i; + ADP	Slow

* Modified from Chance and Hollunger (1963c).

Some of the respiratory rates are obvious; for example, under conditions in which substrate or oxygen is exhausted (States 2 and 5), respiration is markedly slowed. Respiratory inhibitors, such as cyanide, carbon monoxide, amytal, or others, when added in excess to mitochondria respiring in any metabolic state (not shown in Table 5–1) slow or stop respiration. However, when substrate or a pure respiratory inhibitor is present but is not in excess, respiration proceeds at a rate less than usual for the particular metabolic state, but at a normal efficiency. Uncouplers affect both respiratory rate and efficiency, as will be discussed presently.

Hackenbrock's (1966) observations on correlated changes between the internal structure and the metabolic state of mitochondria represent an advance in our understanding of the relationship between function and structure. When the metabolic state changes from 4 to 3, with an acceleration of respiration, there is a rapid reversible condensation of electron-dense 60 to 100 Å granules in the mitochondrial matrix, indicating a rearrangement of the inner mitochondrial matrix in connection with the functional process.

The phenomenon of *respiratory control* by ADP + P_i (States 3 and 4) is the most important feature of mitochondrial metabolism for understanding the physiology of energy-transfer. It adds a dimension of control to the function of the mitochondrion. In a solution of free enzymes, energy can be transformed as efficiently as in the mitochondrion, and oxidation will continue as long as there is substrate available. Like a fire consuming as much wood as is available, the liberation of energy slows somewhat as ash accumulates; however, such a mass-action control is not very useful for maintaining other than British living rooms at a reasonably constant temperature. Mitochondrial respiratory control provides a more modern mode of energy liberation, damped and slowed by the demand for energy, although the fuel is readily available for burning when the demand arises.

Respiratory control is readily demonstrable in isolated intact mitochondria. It is also demonstrable in intact whole cells (Chap. 6), perfused organs (Chap. 6), and the intact mammal (Chap. 7).

How do mitochondria control their respiration through the amount of $\sim$P-bonds formed? The answer to that question must depend upon hypothesis. There are at present three general hypotheses on mechanisms for the transfer of energy to form energy-rich bonds. They differ according to which type of intermediate energy-rich state they postulate in the mitochondrial apparatus: (1) a chemical bond, (2) an accumulation of hydrogen ions and electrochemical potential, or (3) a change in the conformation of a macromolecule. The first two of the theories of oxidative phosphorylation are capable of accommodating respiratory control; indeed, they were designed in part to accommodate it.

The *chemical* theory postulates that oxidative phosphorylation proceeds via the formation of chemical intermediates; for example, Slater (1966a) postulates

Equation 5–2 $\qquad AH_2 + B + C \rightleftharpoons A \sim C + BH_2$

where AH_2 is a reduced intermediate, B an oxidized intermediate, and C yet another intermediate molecule. A and B may be identified with electron-carriers in the chain. C, however, is a special intermediate. For one thing, it forms an energy-rich complex with A via the energy difference between AH_2 and B. That energy-rich potential can be transferred without any external energy source:

Equation 5–3 $\qquad A \sim C + ADP + P_i \rightleftharpoons A + C + ATP$

Equations 5–2 and 5–3 account for the observation that oxidation and phosphorylation are coupled (in the "trivial" sense) by postulating that both processes share a common component, $A \sim C$; if oxidation does not produce $A \sim C$, phosphorylation cannot proceed because of the lack of the necessary energy-rich intermediate. Equations 5–2 and 5–3 can also rationalize the phenomenon of respiratory control or tight coupling, if we postulate that the component C is present in amounts that are very low compared with the amount of the electron-carriers A and B. Under this condition, when $ADP + P_i$ is omitted and phosphorylation does not proceed, the intermediate $A \sim C$ accumulates, because it is not transformed to $A + C$; since respiration requires free C, respiration will slow down or stop (State 4). Adding $ADP + P_i$ will liberate free C (by forming ATP), which will accelerate respiration (State 3).

The equations 5–2 and 5–3 are used as summaries of either substrate-linked phosphorylation or respiratory-chain phosphorylation. Substrate-linked phosphorylation occurs via the interactions of mitochondrial dehydrogenases and kinases and has been described in some detail in Chapter 2. An aldehyde, pyruvate or α-ketoglutarate, is represented by AH_2. These phosphorylations have the common features in that a nonphosphorylated intermediate is formed before P_i or ADP enters the reactions, in that P_i reacts before ADP, and in that C in Equation 5–2 represents an —SH group on the specific dehydrogenase. Through substrate-linked phosphorylation, the complex pyruvate and α-ketoglutarate dehydrogenase systems in mitochondria evolve one mole of $\sim P$ per mole of substrate oxidized, plus one mole of NADH (represented by B in Eq. 5–2) which then enters the respiratory-chain phosphorylation system.

Substrate-linked phosphorylations are not controlled by the metabolic state of the mitochondrion in exactly the same way as in respiratory-chain phosphorylation: uncoupling agents do not discharge the substrate-enzyme high-energy intermediates nor does oligomycin inhibit the respiration that supports the phosphorylation. But the rate of substrate oxidation is a limiting factor, and is controlled by the respiratory chain, probably through the amount of NAD^+ available. Uncoupling agents do accelerate substrate-linked phosphorylations by raising the rate of electron-transport chain respiration (State 3u). The rapid respiration induced by DNP may raise the amount of substrate-linked phosphorylation (one $\sim P$ per mole of substrate) so markedly that it compensates for the DNP-induced loss of respiratory-chain phosphorylation (three $\sim P$ per mole of substrate), and a normal amount of $\sim P$ is evolved per unit of time even though the efficiency (the P:O ratio) is depressed (Slater and Hülsmann, 1959).

The reactions shown in Equations 2–5 and 3–5 are almost certainly incorrect for several reasons. They are shown as if A, B, and C were free in solution. If AH_2 and B represent components of the respiratory chain, they are bound to a membrane and are restricted in motion and diffusion thereby and are probably fixed in position toward each other. The representation of such oriented components of the reaction is more complex than the simple chemical notations used here, but it should be understood that that is what is meant.

With that understanding, a second objection is partly answered; the forward oxidative reaction and the phosphorylative reactions in both directions would be highly improbable chemically if they were taken to imply, as they seem to, that three components free in solution combine to produce a chemical change. Kinetically, the changes of such third-order interactions are small. Since the components are not free but are oriented, the kinetic objection is partly overcome. A further resolution to this difficulty has been to postulate that Equations 5–2 and 5–3 are overall summaries of a series of partial and simpler reactions. Such a resolution, however, requires that one postulate yet a greater number of hypothetical high-energy intermediates; for example, Equation 5–3 might involve another nonphosphorylated intermediate and some phosphorylated intermediates:

Equation 5–4
$$A \sim C + D \rightleftarrows A + C \sim D$$
$$C \sim D + P_i \rightleftarrows C + D + ATP$$

Such formulations have been invoked to explain several experimental findings that indicate that DNP might act on a high-energy intermediate closer to the respiratory chain than the one that arsenate or oligomycin acts upon.

Another possibility is the formation of high-energy phosphorylated intermediates. Complexes like:

Equation 5–5
$$A \sim C + P_i \rightleftarrows A + C \sim P$$
$$C \sim P + D \rightleftarrows C + D \sim P$$
$$D \sim P + ADP \rightleftarrows D + ATP$$

have been invoked as steps in the phosphorylation reactions to explain certain features of exchange reactions between phosphorus atoms (see Slater, 1966a, for a brief summary).

Lastly, the oxidation reaction in Equation 5–2 shows a high-energy complex between the oxidized form of an electron-carrier $(A \sim C)$, and some workers have evidence that the complex is with the reduced form, $BH_2 \sim C$.

The high-energy intermediates have been invoked as the source for certain cellular processes that require energy but do not use ATP directly. Among these processes are the energy-linked pyridine nucleotide transhydrogenation, the reversal of anaerobic electron-transport, and the translocation of small ions—all occurring in mitochondria; they utilize nonphosphorylated intermediates. Certain syntheses may also involve energy-linked mechanisms. ATP does not appear to support the mitochondrial incorporation of amino acids into protein as shown by the failure of oligomycin (which blocks ATP formation) to inhibit the incorporation; high-energy intermediates are suggested as the energy source (Bronk, 1963). In fat cells, protein synthesis apparently subsists on a high-energy source formed from ATP, not an intermediate in ATP formation (Jarrett, 1967). The mitochondrial factor (Sokoloff's Factor) that accelerates ribosomal translation and is produced through an action of thyroid hormones may involve similar processes (Hoch, unpublished data).

Iodohistidine and phosphoriodohistidine have been proposed as intermediates in oxidative phosphorylation in beef heart mitochondria (Perlgut and Wainio, 1964, 1966); however, Holloway *et al.* (1967) could identify no iodohistidine in beef heart mitochondria.

Whatever the details of the chemical hypothesis of respiratory chain oxidative phosphorylation may turn out to be, it is clear even now that the hypothesis is complex and needs to have new and hypothetical intermediates interposed between the earlier postulated intermediates as new experimental data are evolved.

The *chemiosmotic* theory has been advanced in recent years mainly because, as appealing as the chemical theory is, the postulated chemical intermediates are not yet identified after some 20 years of intense work. This defect may be merely a technical problem, imposed by the transient nature or trace concentrations of the chemical intermediates, but one must also consider the possibility that the chemical intermediates just do not exist.

The chemiosmotic theory proposes that in or on the mitochondrion is a membrane that is impermeable to protons and that contains an enzyme that catalyzes the reaction

Equation 5–6 $ATP + H^+ + OH^- \rightleftarrows ADP + POH$

in a vectorial manner. The enzyme is oriented so that in one direction across the membrane it acts only to catalyze the forward reaction in Equation 5–6, i.e., as an ATPase. In the other direction across the membrane, the enzyme catalyzes only the back reaction, i.e., as an ATP synthetase.

The chemiosmotic model shows "trivial" coupling as depending upon the formation of a gradient of H^+ and electrochemical potential inside the mitochondrion via the production of H^+ ions inside the membrane when electrons from oxidation of substrate (SH_2) are transported to the outside phase, where they combine with O_2 and H^+: effectively an H^+-transfer to the inside (see Fig. 5–2). Such a gradient has actually been measured, and in normal rat liver mitochondria amounts to about 230 mv; Mitchell calls this the "proton-motive force." The "inside" H^+ gradient drives the anisotropic ATPase in reverse to achieve a net H^+-transfer to the outside, and to synthesize ATP from $P_i + ADP$. The respiratory coupling—i.e., the slowing of respiration (State 4)—is a result of the accumulation of H^+

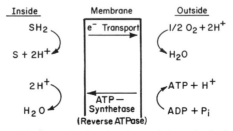

Figure 5–2. A general scheme for the chemiosmotic hypothesis of oxidative phosphorylation.

inside when oxidation continues in the absence of ADP $+ P_i$; the heightened proton-motive force retards the oxidative cycle. Respiratory control is the result of competition for H^+ pumping between the anisotropic ATPase and electron-transport. These postulates are greatly strengthened by Mitchell's experiments showing the amount and rate of H^+-transfer in mitochondria (Mitchell and Moyle, 1969).

However, there is still considerable disagreement about both the data on H^+ transfer in mitochondria, and the interpretation of observations that H^+ and other cations are actually transferred. While Mitchell and his supporters cite such transfers as the driving forces for oxidative phosphorylation, others say they depend upon energy transformed via oxidative phosphorylation and are effects rather than causes. Chance *et al.* (1967 b) point out that measurements of ATP formation in response to externally impressed pH gradients (one of the methods used to test the chemiosmotic theory) cannot decide whether energy is conserved via chemical intermediates or proton (and EMF) gradients; they also summarize kinetic evidence inconsistent with chemiosmosis. The chemiosmotic hypothesis is more generally accepted for the function of chloroplasts. The interest in this area makes it likely that crucial measurements in the next few years will test the validity of both the chemiosmotic and chemical hypotheses. Already both have been modified to accommodate recent findings.

The *conformation* theory postulates that oxidative phosphorylation proceeds via the formation of energy-rich conformations of macromolecules (see Chance *et al.*, 1967; Penniston *et al.*, 1968; Harris *et al.*, 1968b; Green *et al.*, 1968; Green and Baum, 1970). In place of chemical intermediates or a proton-motive force, the changes in the free energy of oxidation alter the tertiary or quaternary structure of a protein or a membrane. Green considers that the energized conformational states of mitochondria are the functional equivalents of illusory high-energy chemical intermediates (Fig. 5–3). Although the original formulations of this theory are not completely explicit, the changes in conformation that an electron-carrier might undergo between an oxidized and a reduced state offer an interesting basis for this sort of thinking.

The most certain evidence for differing conformations is now beginning

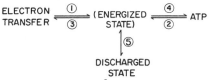

Figure 5–3. The generation and discharge of the energized state of mitochondria. Both "energized" and "energized-twisted" configurations of the *cristae mitochondriales* are called "energized" in this diagram. Substrates (1) or ATP (2) can generate the energized configuration, and reversed electron-transport (3) or ADP (4) discharge it by reverse reactions. Other modes of discharge (5) are through the action of uncoupling agents, or by ion-induced swelling, or by the translocation of divalent metal ions. (After Green *et al.*, 1968.)

to appear from the determinations of the three-dimensional structure of electron-carriers by x-ray crystallography, but there has been less direct evidence that depends upon changes in chemical and physical properties. Hemoglobin is the best example to date, but it, of course, is not an electron-carrier; the hemoglobin structure is rearranged when the heme iron binds or loses oxygen. The conformation of ferro-cytochrome *c* appears to differ from ferri-cytochrome *c*, as shown by general alterations in the crystal state (the specific structure of the ferro form is not yet reported) and in optical and chemical behavior (Dickerson *et al.*, 1968). The conformation of some flavin enzymes and carriers also may depend on the redox state of the flavin, but only crystals of soluble enzymes have been studied by x-ray methods so far, and the isolation of the flavin members of the electron-transport chain is still not completed. Among the pyridine nucleotide-dependent enzymes, the glutamate dehydrogenase of liver depolymerizes from a molecular weight of 1,000,000 to at least four subunits in the presence of NADH or metal-binding agents. Boyer (1964) had suggested that changes in con-formation might be a basis for energy conservation, and early studies of Lehninger (1960) on "mechano-enzymes" and the work of Weinbach and Garbus (1969) on uncoupling agents contribute supporting evidence.

Further support for the conformation theory comes from the electron micrographic studies of mitochondria by Hackenbrock (1966, 1968, and see p. 46), in which the electron-density of the mitochondrion is shown to vary with the metabolic state. Such alterations probably involve the rearrange-ment of structures much larger than the individual carrier molecules like cytochrome *c*. The current experimentation on the reconstitution of mito-chondrial function from soluble and membranous fractions may be expected to give further crucial information. The conformation theory is more akin to the chemical intermediate theory than to the chemiosmotic theory; there is little intellectual difficulty in accepting an alteration in conformation of a protein when it complexes a smaller molecule or another protein.

However, difficulty does arise when one asks whether a new confor-mation of a molecule can account for the 7 kcal per mole necessary to evolve an energy-rich bond, as would be required in oxidative phosphorylation according to a conformation theory. There is as yet no evidence that changes in the electron-carriers can account quantitatively for such an energy dif-ference, but, again, studies on hemoglobin indicate a surprisingly large value. The difference between oxygenated and deoxygenated hemoglobin, cal-culated from the sum of a number of low-energy intramolecular interactions, is about 8 kcal per mole (Wyman, 1964; Noble, 1969). This large total energy difference is distributed over the interacting surfaces of the four subunits of hemoglobin, and it remains to be shown that it can be con-centrated into the formation of a single energy-rich bond. In addition to having to account for the necessary amount of energy, a kinetic requirement also exists: are the changes in conformation of electron-carriers fast enough, even if they may be big enough? Azzi *et al.* (1969) present evidence that structural changes in mitochondrial membranes that are associated with energy conservation occur at a time rate too slow to account for electron-transport; this suggests they are effects rather than causes.

UNCOUPLING: LOSS OF RESPIRATORY CONTROL AND EFFICIENCY

The phenomenon of *uncoupling* refers to the nontrivial sense of the relation between oxidation and phosphorylation. *Oxidation proceeds without phosphorylation.* (The converse, phosphorylation proceeding without oxidation, is unthinkable.) Various experimental means produce uncoupling. Damage to the structure (membrane?) of mitochondria by hypotonic

Figure 5–4.   The structure of 2,4-dinitrophenol (DNP).

solutions, freezing and thawing, ultrasonic vibrations, grinding or heating all produce mitochondria that oxidize but do not phosphorylate. Chemical agents by the dozens have been shown to uncouple; the classical one is 2,4-dinitrophenol, first used by Loomis and Lipmann (1948) and Cross *et al.* (1949).

Not only does uncoupling decrease the efficiency of phosphorylation, but it also decreases the degree of respiratory control (Fig. 5–5). Uncoupled

Figure 5–5.   Same experiment as in Figure 5–1, but with the further addition of DNP; respiration is accelerated by DNP (State 3u) more than by ADP (State 3) in this instance.

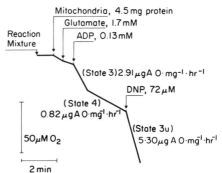

mitochondria respire at a high rate (State 3u, see Table 5–1). Long before the mechanism was known, respiratory stimulation of cellular respiration by DNP was observed (DeMeio and Barron, 1934; von Euler, 1932; Tainter and Cutting, 1933).

These observations are also accommodated by the chemical and chemiosmotic theories of oxidative phosphorylation. By the chemical theory, it is postulated that DNP acts (somehow) to hydrolyze the hypothetical energy-rich intermediate:

Equation 5–7 $A \sim C \xrightarrow{\text{DNP}} A + C$

The consequences of this discharge of $A \sim C$ are two-fold: with no $A \sim C$ being formed, Equation 5—3 will be forced from left to right, i.e., as an ATPase; and with free C being supplied rapidly through the reaction in Equation 5–7, Equation 5–2 will proceed more rapidly from left to right, i.e., respiration will be maximal. Both phenomena are observed: the ATPase activity of mitochondria is activated by DNP, and respiration is stimulated (State 3u). The experimental evidence for Equation 5–7 will require the demonstration of a nonphosphorylated energy-rich complex that is hydrolyzed by DNP or by a DNP-activated enzyme.

The chemiosmotic theory does as well or better. It is postulated that DNP enters the lipid-containing membrane, being lipid-soluble, and there produces a "proton-leak," perhaps by the DNP phenolic group conducting H^+ ions. With the discharge of the H^+ gradient that existed inside the mitochondrion, respiration can proceed at its maximal rate, but the proton-motive force driving the ATP synthesis is short-circuited and only the DNP-activated ATPase is observed. These postulates are much strengthened by observations that DNP in uncoupling concentrations (50 μM) does make mitochondria discharge H^+ ions and lowers the proton-motive force (Mitchell and Moyle, 1969).

The conformation theory has a place in it for uncoupling. The action of DNP on mitochondria has many aspects that would be explained by a change in the conformation of proteins, such as increases in mitochondrial permeability and in susceptibility to proteolytic enzymes (Weinbach and Garbus, 1968) and in the kinetics of respiratory activation fitting the Hill equation (Hoch, 1968b). Electron micrographs show an ultrastructural dislocation in the cristael membranes of beef heart mitochondria that have been uncoupled by lypohilization, and it is suggested that uncouplers induce a low-energy configurational state in such membranes (Jolly *et al.*, 1969).

LOOSE-COUPLING: LOSS OF RESPIRATORY CONTROL WITH NORMAL EFFICIENCY

The phenomenon of loose-coupling refers only to mitochondrial respiration: *respiration proceeds at a more rapid rate, but phosphorylation is normal.* Calling this loose-coupling is perhaps misleading, because coupling can refer either to phosphorylation or to respiration.

Low concentrations of agents capable of producing uncoupling accelerate mitochondrial respiration while phosphorylation is still at normal or almost normal efficiency. Thus, DNP at concentrations between 1 μM and

10 μM stimulates respiration up to three-fold, but decreases the efficiency of phosphorylation only up to 25 per cent. To demonstrate the phenomenon of loose-coupling, both respiration and phosphorylation must be measured directly; the indirect estimation of an ADP:O ratio from the transition between State 3 and State 4 does not differentiate between uncoupling and loose-coupling, because in both there is no respiratory control.

Loose-coupling fits into the two theories of oxidative phosphorylation with about equal ease. In the chemical theory, we may postulate that loose-coupling is due to the action of an uncoupling agent that discharges C at a less than maximal rate (Eq. 5–7). The rate of C liberation is low enough so that phosphorylation (Eq. 5–3) competes successfully for A $\sim$ C and continues at nearly maximal efficiency, whereas the oxidative reaction (Eq. 5–2) is supplied with enough free C to enable it to proceed at nearly maximal rates. In other words, we must postulate that oxidation and phosphorylation have different affinities for C. Mitchell warns that "the building of second order hypothesis upon first order hypothesis, exemplified here, and not uncommon in the literature of oxidative and photosynthetic phosphorylation, is known to be a rather hazardous practice." Direct measurements of free C are not yet possible.

The chemiosmotic theory also accommodates loose-coupling. There, we must postulate a similar difference in the degree of dependence of the rates of oxidation and phosphorylation upon the proton-motive force built up inside the mitochondrial membrane. Mitchell and Moyle (1969) have shown experimentally that such a difference exists. When enough DNP is added to rat liver mitochondria to depress the proton-motive force from its normal value of 230 mv to about 200 mv, oxidation becomes rapid but phosphorylation still proceeds at nearly normal levels.

Loose-coupling is of interest in considering the physiologic state of the resting cell, and the control mechanisms that regulate oxidative rates. It will be shown in Chapter 6 that resting respiration in the whole cell, in organs, and in the intact organism proceeds in State 4 as if little or no ADP + P_i were available to the mitochondrial systems. Respiration in State 4 generates nonphosphorylated energy-rich bonds that can support, for instance, intramitochondrial ion-pumping. Any agent, endogenous or exogenous, that can stimulate resting respiration without destroying phosphorylation will increase the output of high-energy bonds in State 4.

INHIBITORS OF PHOSPHORYLATING RESPIRATION

This group includes a number of complex compounds with antibiotic properties, usually isolated from living lower organisms; oligomycin is the best studied. Others which act in many, but not all, respects like oligomycin are aurovertin, valinomycin, atractylate, rutamycin, triethyltin, guanidine, and so forth (Fig. 5–6).

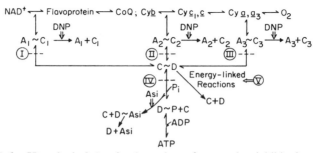

Figure 5-6. Hypothetical sites for the actions of agents that inhibit electron-transport in State 3 (i.e., the accelerated respiration that occurs in the presence of ADP + P_i), and some uncoupling agents. The components $C_{1,2,3}$ are necessary for electron-transport at the sites shown; the reagents in groups I, II, and III inhibit the release of $C_{1,2,3}$, and those in IV and V act at a common site. Uncoupling agents like DNP relieve these inhibitions by liberating $C_{1,2,3}$ in energy-wasting reactions; arsenate acts similarly by competing with P_i; gramicidin and valinomycin uncouple in the presence of alkali metal ions. Representative members of the groups are:

(I)   4-Methyl, 3-butene (guanidine); hexyl- and octylguanidine; amytal.
(II)   Phenethylbiguanidine; n-Heptylquinoline-N-oxide.
(III) Decamethylene diguanidine.
(IV)   Oligomycin; aurovertin; valinomycin; triethyltin.
(V)   Gramicidin; valinomycin.

These inhibitors depress the rate of mitochondrial electron-transport only when phosphorylation is proceding. They act on State 3 respiration in the presence of excess substrate, ADP, and P_i (see Eqs. 5–3 and 5–7) to inhibit the transfer of energy at a step involving the formation of phosphorylated intermediates. To explain their action in terms of chemical intermediates, these inhibitors prevent the liberation of C to support oxidation (Eq. 5–2), and so respiration is slowed (or in the chemiosmotic theory by allowing H^+ ions to leak out and thereby overcoming the H^+ back-pressure owing to the oligomycin inhibition of the ATPase). If C is supplied via another reaction, as in Equation 5–7, in which an uncoupling agent promotes the hydrolysis of $A \sim C$, the inhibitors of phosphorylating respiration do not depress oxidation; they do not inhibit respiration in State 3u. Similarly, when phosphorylation is not operative because of the absence of ADP + P_i and respiration proceeds slowly because little free C is supplied, these inhibitors have no effect; they do not inhibit respiration in State 4. The action of this group of inhibitors seems to be reasonably specific—in most cases, when they do inhibit respiration, that respiration is coupled to phosphorylation, and when they fail to inhibit respiration, that respiration is not coupled to phosphorylation. Oligomycin and the other agents have been used extensively in studies on the mechanism of oxidative phosphorylation to test for phosphorylating processes. Indeed, the inhibition by oligomycin is one of the three most important characteristics of oxidative phosphorylation (the other two are respiratory control and uncoupling by dinitrophenol) (Slater, 1966a).

Olygomycin inhibits not only respiration, but also some of the partial reactions of oxidative phosphorylation that are implied in Equations 5–3 and 5–5: the exchange between P_i and ATP, the exchange of oxygen between Pi and water, and the DNP-induced ATPase. These findings are the basis for believing that oligomycin acts on a phosphotransferase reaction.

ALLOSTERIC CONTROLS

The pathways of energy transformations not only are controlled by the availability of ADP, but are regulated by the same sort of feedback mechanisms that act upon other metabolic processes. Substances evolved during metabolism may be positive or negative modifiers of the action of enzymes that catalyze earlier steps in the sequence of reactions. These modifications appear to involve allosteric mechanisms, whereby the modifier alters the specific activity of the enzyme by changing its structure at sites distinct from the active catalytic center. These modifiers are noncompetitive inhibitors or activators, in contrast to competitive agents that act at the catalytic site, where substrates also react. It will be recognized that ADP may control the rate of mitochondrial electron-transport through a similar allosteric mechanism.

The glycolytic and oxidative pathways of carbon metabolism that lead to the formation of NADH are subject to feedback allosteric control at several crucial enzymatic steps. Among the modifiers are the ultimate products of energy-transfer: ATP, ADP, AMP, and P_i (Fig. 5–7). Other modifiers act on these pathways, and the synthesis of some, like acetyl coenzyme A, depends directly on energy transfers from ATP.

The allosteric regulation of glycolysis and of the Krebs cycle appears to be purposive. The accumulation of chemical potential in the form of ATP acts to slow down the further synthesis of ATP, and the dissipation of ATP

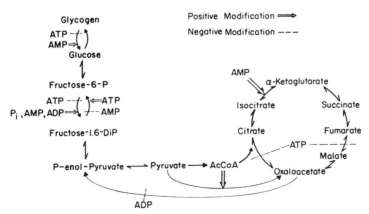

Figure 5–7. Some allosteric controls of energy transformations through the actions of ATP, ADP, AMP, P_i, and acetyl coenzyme A (AcCoA) as positive or negative modifiers.

accelerates the synthesis of ATP. The combination of a high concentration of ATP and low concentrations of ADP, AMP, and P_i facilitates the formation of glucose-6-phosphate by inhibiting the phosphofructokinase enzyme and activating the fructose-1,6-diphosphate phosphatase; the supply of oxidizable substrate is thereby diminished. The flux of 2-carbon fragments in the Krebs cycle is slowed, and less NADH is supplied to the electron-transport system. Conversely, when the concentration of ATP is decreased, as by utilization or starvation, more acetyl coenzyme A is made available to an accelerated Krebs cycle.

AMP plays a special role as an enzyme modifier, especially on the isocitrate dehydrogenase and the glycogen phosphorylase. AMP is formed from ATP by several important processes, including amino acid activation and fatty acid activation, which are accompanied by the formation of pyrophosphate (PP_i):

Equation 5–8 $$ATP \rightarrow AMP + PP_i$$

AMP also serves in the resynthesis of ATP through the action of the *adenylate kinase*:

Equation 5–9 $$AMP + ATP \rightleftharpoons 2\ ADP$$

This enzyme is widely distributed in tissues and catalyzes a dismutation reaction, transferring a $\sim$P-bond from ATP to AMP; the ADP so formed can be further phosphorylated by mitochondrial oxidative phosphorylation. In the reverse direction, the adenylate kinase reactions maximize the utilization of the $\sim$P-bonds of ATP, e.g., muscle contraction utilizes only the first $\sim$P-bond of ATP and produces ADP. When the utilization of ATP so predominates over its synthesis that ADP accumulates and more energy is needed, the adenylate kinase (called, in muscle, *myokinase*) makes available the second $\sim$P-bond and produces AMP.

Studies on the metabolic pathways of bacteria have lead to a general formulation of the energy status of the cell as a regulatory parameter, as measured by the relative concentrations of ATP, ADP, and AMP. Atkinson (1968a) defines a parameter as the "energy charge":

Equation 5–10

$$\text{Energy Charge} = \frac{[ATP] + 0.5\,[ADP]}{[ATP] + [ADP] + [AMP]}$$

and presents experiments on *Escherichia coli* that support the concept that a high-energy charge decelerates ATP—regenerating enzyme sequences, and accelerates ATP-utilizing sequences; a low energy charge has the opposite effect. Similar alterations in *Salmonella typhimurium* suggest that the energy status regulates protein synthesis at the activation step; ADP and AMP inhibit several amino acyl-tRNA-synthetases (Brenner *et al.*, 1970). A low energy charge thus slows a major ATP-utilizing system by a direct action on an enzyme, as well as through the general limitation of the availability of ATP as a component of energy-requiring reactions. It will be interesting to see if these regulatory mechanisms occur in mammalian systems.

General References: A collection of papers tracing the development of concepts of biological phosphorylations has been compiled by Kalckar (1969). The chemical theory of oxidative phosphorylation is reviewed by Slater (1966a, b); the chemi-osmotic theory by Mitchell (1966, 1968); and the conformation theory by a series of papers from Green's laboratory (1966, 1968, and Green and Baum, 1970). Williams (1969) discusses electron-transport and energy conservation in terms of chemical physics. Greville (1969) evaluates dispassionately the apparent conflicts between chemiosmotic and chemical-intermediate protagonists. The action of inhibitors of phosphorylating respiration is reviewed by Lardy *et al.* (1958). A theoretical basis for uncoupling is presented by Guillory (1969). Atkinson (1968a, b) discusses the allosteric control of energy transformations. Slater (1967) gives experimental details for the use of inhibitors of phosphorylation and respiration.

CHAPTER

6

ENERGY TRANSFORMATIONS IN THE LIVING CELL

The studies outlined in the previous chapters indicate how mitochondria function when they are isolated from the cell and are placed in an artificial medium. Such information is valuable, of course. It tells us what mitochondria are capable of doing. Presumably, they will perform in the living cell at least as well and as completely as they do *in vitro*, so the *in vitro* phenomena are a starting point for understanding behavior *in vivo*. In this sense, we are in the same situation as most scientists in trying to understand living systems through a reduction hypothesis: one assumes the behavior of an isolated system gives information (admittedly incomplete) on the behavior of that system when integrated into a more complex set of interdependently functioning systems.

However, we may ask not only "what *can* mitochondria do?" but also "what *do* mitochondria do?" with some promise of a reasonable answer at this time. This promise arises from the accessibility of mitochondria to experiment even when they are in the living cell. The mitochondrial content of pigments that absorb specific wavelengths of light, and the development of chemical agents that are reasonably specific for mitochondrial and other metabolic functions, have afforded experimental routes for gaining information on mitochondrial behavior *in vivo*. Figures 3–5 and 5–6 summarize some of the inhibitors and other agents used in such metabolic studies.

In general, studies on energy metabolism *in vivo* may fall into several categories, according to the complexity of the biologic structures that are being examined. Separated living cells, tissue slices, whole organs, and live animals have been examined. It may be stretching the definition of "*in vivo*" somewhat to include cells and slices, but it will be assumed here that they are more "intact" and offer a more natural environment to the mitochondria than the synthetic reaction mixtures used *in vitro*.

SINGLE CELLS

The problem of cell impermeability limits the interpretation of experiments on energy transformations in large populations of intact cells, whether in suspension or in slices or whole organs. This problem has been approached by studies on single large cells, using microscopy and sensitive spectrophotometric or fluorimetric measurements of redox states and kinetics (Chance and Legallais, 1959; Chance *et al.*, 1959; Chance *et al.*, 1967a; Kohen *et al.*, 1966, 1968). The cells, obtained from ascites tumor cell cultures after x-irradiation, are called EL-2 "giant" cells and measure about 80 by 20 μ. Metabolites are introduced through microinjection and microelectrophoresis, and portions of a single cell are examined through a 10-μ aperture in the microfluorimeter, which can observe events in the cytosol, nucleus, or mitochondria.

Injection of substrates of the glycolytic pathway into the cytosol, especially fructose-1,6-diphosphate, together with ADP $+$ P$_i$, reduces pyridine nucleotides in the cytosol rapidly, as expected. Among the substrates of the Krebs cycle, malate, α-ketoglutarate, succinate $+$ ATP, and glutamate reduce pyridine nucleotides in the mitochondrial space; malate alone causes considerable reduction of pyridine nucleotides in the cytoplasm, demonstrating the existence of a malate dehydrogenase in the cytoplasm, and its potential control over the extramitochondrial redox state.

CELL SUSPENSIONS

Energy transformations in cell suspensions are made difficult not only by the selective permeability of the cell wall to phosphorylated compounds such as ATP but also by problems in the preparation of suspensions of separate cells from organs, arising from the adhesiveness of cell walls. Ingenious circumventions of these complexities have, however, been devised.

Progress in tissue culture techniques has made possible the preparation of dispersed cells from livers (Ichihara *et al.*, 1965). In the hands of these workers, the endogenous respiration of liver cell suspensions (20 to 30 mg dry weight) in a Ca^{2+}-free Locke's solution is about 4 μmoles of O_2 per minute. Adding succinate as a substrate increases respiration about six-fold. These whole cells oxidize added succinate with a relatively good degree of respiratory control. They also oxidize other substrates normally serving as sources of reducing equivalents for mitochondrial electron-transport, such as glutamate, malate, and pyruvate, with similar degrees of respiratory control, and the rates relative to those seen with succinate are similar to the rates observed in isolated mitochondria. Added DNP markedly increases respiration with added succinate, and then ADP does not stimulate respiration further. This is typical uncoupling behavior; however, these liver cells appear to have lost glycolytic enzymes and cofactors, since their capacity for glycolysis and gluconeogenesis is limited.

Respiratory control is modified in these whole cells (Fig 6–1). Whole cells respire with a more gradual transition between State 3 and State 4 than is shown by mitochondria isolated from the cells. Apparently the cytoplasmic components surrounding the mitochondria hydrolyze or utilize the ATP formed in State 3 to produce ADP, and the difference in the regulatory patterns represents the competitive effects of oxidative phosphorylation and ATPase activity. P:O ratios of 2 are observed with pyruvate in whole cells, which compares well with the P:O ratios observed in the mitochondria.

It thus appears that mitochondrial oxidative metabolism and regulation proceed in whole liver cells much as they proceed in isolated liver mitochondria, with some explicable modifications due to the different environments. Further, the oxidation of added substrates by the intact liver cells proceeds in the controlled metabolic state (State 4), as shown by the stimulatory effects of ADP. The importance of this observation will be discussed

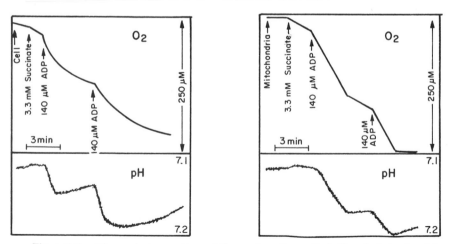

Figure 6–1. The respiratory pattern of dispersed rat-liver cells (left) and of mitochondria isolated from the dispersed cells (right), assayed polarographically under similar conditions. (From Ichihara *et al.*, 1965, by permission of the authors and Elsevier Publishing Company, Amsterdam.)

in Chapter 7, when the oxygen consumption of the whole animal is considered.

Among the normal cells that have been examined, muscle cells have special features. Actomyosin is a ready-made *in vivo* phosphate-acceptor system. Stimulated individual skeletal muscles or isolated fibers can be made to contract, or myocardial cells can be cultured and then they contract rhythmically and spontaneously. Since muscle contraction or relaxation depends directly upon ATP, one can study correlations between muscle mitochondria *in situ* and the energy made available by oxidation.

The rate of spontaneous beating of single heart cells in culture is affected by oligomycin, DNP, and ouabain, and appears to be supported by ATP, produced either by glycolysis or mitochondrial oxidative phosphorylation (Harary and Slater, 1965). (The measurement of the rate, rather than the force of contraction, may perhaps leave these studies open to the interpretation that the rate-controlling mechanisms that impart periodicity depend upon an ATP supply of questionable stoichiometry rather than that actomyosin acts stoichiometrically to function as an ATPase.) Adding DNP slows the beating rate, and then adding ATP or oligomycin restores the rate; these findings are interpreted to mean that DNP uncouples oxidative phosphorylation and stimulates the mitochondrial ATPase, depleting the ATP supply from oxidative phosphorylation and hydrolyzing the ATP supply from glycolysis. Oligomycin acts under such conditions to inhibit the mitochondrial ATPase, and although oligomycin also inhibits the mitochondrial phosphorylating respiration, glycolysis is able to supply ATP to support beating. The addition of iodoacetate to this system stops beating by inhibiting glycolysis. The high-energy nonphosphorylated intermediates formed in mitochondria before the site of oligomycin inhibition obviously are not able to support muscle contraction.

Tight-coupling of oxidative phosphorylation in suspension of intact cells can be demonstrated through the use of oligomycin, which does not inhibit uncoupled respiration. In cultured mammalian cells, when precautions are taken in handling to avoid inducing loose-coupling, oligomycin is an extremely potent and stoichiometric inhibitor of endogenous respiration (Currie and Gregg, 1965; Gregg *et al.*, 1968). Again, the kinetic behavior observed differs from that seen with mitochondria isolated from such cells. In the intact cells titrated with oligomycin, the first addition produces respiratory inhibition and each subsequent addition produces a linear increase of inhibition of respiration. In the isolated mitochondria, the initial additions of oligomycin do not inhibit until a threshold level is reached, and only then do subsequent additions inhibit the rate of respiration linearly. The endogenous respiration of cultured ascites tumor cells is also tightly coupled, as shown by the observation that oligomycin inhibits respiration up to 85 per cent (Dallner and Ernster, 1962).

Whole cell suspensions are prepared more easily when certain strains of tumor cells are grown than when normal organs are disintegrated, an advantage that must be weighed against the fact that tumor cells may have rather different metabolic patterns and control mechanisms than do normal cells. Although the respiration of tumor cells is inhibited strongly by oligomycin, there is also evidence that normal cells and tumor cells have different patterns for energy metabolism and utilization. DNP (10 μM) and oligomycin (0.025 μg per mg protein) each inhibit protein synthesis in anaerobic thymic lymphocytes by about 50 per cent, and the two agents together inhibit by about 70 per cent, as if they acted on subsequent steps (Jarret and Kipnis, 1967). The same concentration of oligomycin (0.025 μg per mg protein) inhibits protein synthesis completely in aerobic lymphocytes or isolated fat cells. In contrast, anaerobic ascites tumor cells show no decrease in the incorporation of ^{14}C-leucine into protein in the presence of 200 μM DNP or of oligomycin. It seems that in yeast, bacterial cells, lymphocytes, and fat cells, protein synthesis is coupled to an ATPase that is sensitive to DNP and oligomycin. In ascites tumor cells, this step is lost or altered through a modification that allows the process of protein synthesis to utilize ATP directly rather than via its conversion by a membrane-localized ATPase as in normal cells.

Lipocytes, fat cells from adipose tissue ("brown fat"), have been prepared in suspensions. They pose a special problem for assays of mitochondrial function in that free fatty acids act as uncoupling agents, and spontaneous or induced lipolysis may affect the mitochondria; usually bovine serum albumin is added to eliminate the effects of free fatty acids as much as possible. Under such conditions, oligomycin inhibits lipocyte respiration about 45 per cent which implies that half the respiration of the isolated lipocyte is not obligatorily coupled to phosphorylation or controlled by the concentration of ADP, and therefore produces only heat (Hepp *et al.*, 1968). On the other hand, the effects of the fatty acids *in vitro* and the question of the penetration of oligomycin into the cells make the extrapolation of these results to the situation *in vivo* rather difficult.

Another approach to studying intact cells has been the measurement of redox state of the respiratory pigments through their absorption or emission of light (Chance and Hess, 1959; Chance *et al.*, 1962). Optical methods have the advantages of involving minimal manipulation of cell suspensions and allowing very rapid (milliseconds) measurement of redox states at a number of electron-transport steps, although the limitations of cell wall permeability to substrates, cofactors, and inhibitors are not avoided. The contribution of cell wall permeability can be evaluated in part in ascites tumor cells in such studies by rupturing their cell walls and comparing the function of the isolated mitochondria with the function of the intracellular mitochondria. Yeast cells, however, have required such drastic methods to break their cell walls that the mitochondria were damaged.

Intact ascites tumor cells were shown to contain a complement of mitochondrial electron-transport components that are more than adequate to account for their respiration. There had been some question about this by Warburg (1930), who speculated that the observation that cancer cells continue to glycolyze in the presence of oxygen even while their respiratory rate is high reflects a defect in their respiratory enzymes that is significant in their uncontrolled growth and replication. For instance, measurements of the amounts of cytochrome c or b that were extractable had indicated that ascites cells were deficient in respiratory pigments. Dual-wavelength studies on intact cells indicated this was not the case, suggesting that the extractions must have been incomplete. Measurements of the turnover numbers of the individual cytochromes in ascites cells showed that there was 10 to 40 times as much cytochrome $a + a_3$ as was necessary, i.e., compared with the demands made upon cytochrome oxidase by the O_2-uptake of the resting cell. Earlier workers, e.g., Korr (1939) and Stannard (1939), had also raised the question of shunt or by-pass routes for electron-transport that did not involve mitochondrial enzymes; there have been reports, for instance, that respiration continued in cells in the presence of azide, cyanide, or antimycin a. Chance and Hess were able to inhibit respiration almost completely with amytal or antimycin a and to correlate the inhibitions with the reduction of intramitochondrial pyridine nucleotides. There was no evidence for the existence of significant extramitochondrial by-passes.

In spite of the more than adequate complement of electron-carriers in the mitochondria of ascites tumor cells, the measured rate of electron-transport was about three times less in the intact cells than in mitochondria isolated from the cells (and measured in the presence of excess concentration of substrate, P_i, or phosphate-acceptor). It appeared that in the intact cell a controlling factor depressed electron-transport, but this did not occur in the isolated mitochondria; that factor might be the concentration of substrate, ADP, or P_i. If it were ADP, supplying ADP should accelerate electron-transport about three-fold, up to the level seen in the isolated mitochondria supplied with ADP. These workers introduced ADP into the intracellular mitochondria by adding glucose to the cells; the phosphorylation of the glucose by an intracellular hexokinase used intracellular ATP to form ADP:

Equation 6–1   Glucose $+$ ATP $\rightarrow$ Glucose-6-phosphate $+$ ADP

Upon addition of glucose, respiration almost immediately accelerated about three-fold, indicating that ADP concentration might control mitochondrial respiration in the intact cell rather than P_i concentration or a substrate concentration. That conclusion was strengthened by the reversal of the glucose-induced stimulation of respiration in 2 minutes, when respiration decreased to below the initial level—the well-recognized Crabtree effect. The ADP had been phosphorylated, removing the stimulus to mitochondrial electron-transport.

Thus, the respiratory metabolism of the intact cell is limited by respiratory control, and the low intracellular concentration of ADP appears to be responsible for the low respiratory metabolism of fresh ascites tumor cells.

Phosphorylation has been studied in ascites tumor cells and in bovine spermatozoa by Morton and Lardy (1967). With 2-deoxyglucose as a phosphate acceptor, and fluoride as an inhibitor of the hydrolysis of phosphate compounds and of motility, the P:O ratios were lower than those measured in isolated mitochondria. Modifying the cell membranes by physical means (shaking with glass beads) or by chemical means (surfactants or filipin, a polyene antibiotic), apparently made them selectively permeable to admit P_i, O_2, and added substrates; then the P:O ratios were close to those seen with isolated mitochondria. It will be of interest to see whether these modifications are specific enough to allow the study of mitochondria in the whole cell as readily as isolated mitochondria.

If, as it appears, our understanding of the function of isolated mitochondria may be translated into an understanding of the modified function of mitochondria in the intact cell, the following question may be asked: Does mitochondrial function in the cell account for oxidative energy metabolism in the whole animal?

R. E. Smith (1955–56) has attempted a quantitative answer by correlating the metabolism of liver mitochondria with total body weight and oxygen consumption in mammals. In a series of mammals varying in size from rat to steer, the number of mitochondria per gram of liver decreased and the mass of each mitochondrion increased slightly. The total mass of liver mitochondria (the product of number × mitochondrial mass per gram of liver × total grams of liver) increased with the 0.77-power of the body weight. As will be seen in Chapter 7, total body oxygen consumption in such mammalian arrays increases with the 0.75-power of the body weight. Thus, the amount of oxygen mammals consume is quite well correlated with the amount of mitochondria in their livers. This correlation does not, of course, imply causation. The liver probably consumes up to one-third of the oxygen in resting animals. Similar data on muscle would be of value, inasmuch as muscle is thought to constitute some 40 per cent of the respiring mass of the body.

TISSUE SLICES

The respiration of tissue slices prepared from the organs of rats has been measured as a basis for estimating the contribution each organ system makes to total body respiration (Field *et al.*, 1939). The results are summarized in Table 6–1.

As estimated from the contributions the organ respiration makes to the total oxygen consumption, and taking into regard the weight fraction of that organ per total body mass, this study shows that tissues fall into three degrees of relative activity. As expected, such relatively inert organs as supporting and covering tissues respire little, and have activities much less than 1. Organs like skeletal muscle and heart do not respire much when they are not contracting, and have an activity around 1 (but since heart, at least, contracts under conditions in which the whole animal consumes oxygen, the contribution of an inert heart slice does not really mean much). Those

Table 6–1 *The Per Cent Contribution of Organ Respiration to Total Oxygen Consumption in 150 gm Fasted Rats**

ORGAN	% WEIGHT IN RAT	% CONTRIBUTION TO BMR	SPECIFIC RESPIRATORY ACTIVITY (BMR CONTRIBUTION PER WEIGHT)
Skeletal muscle	41.6	32.7	0.79
Skin	18.5	7.0	0.38
Skeleton cartilage	6.7	0.9	0.13
Blood	6.5	0.15	0.02
Liver	5.5	10.0†	1.82
Gut	4.9	5.3	1.08
Ligament	4.9	0.3	0.06
Brain	0.9	2.6‡	2.89
Kidney	0.9	3.5	3.89
Testis (Ovary)	0.8	0.75	0.94
Lung	0.6	0.7	1.17
Heart	0.5	0.8	1.00
Spleen	0.3	0.3	1.00
Remainder	6.3	1.2	0.19
Not accounted for	1.1	34.2	

* From Field *et al.*, 1939.

† Underestimated. The authors note that liver slices take up water rapidly, so that the specific respiratory activity, measured per gram wet weight of tissue, was too low by a factor of at least 2. Since 34.2 per cent of the tissue contribution to the BMR was unaccounted for, liver respiration might contribute up to 30 per cent or so, even though the tissues not accounted for need low specific respiratory activities.

‡ Underestimated. There was reason for the authors to think that their preparations of brain slices did not respire at capacity.

organs with nonmechanical metabolic functions that continue even *in vitro*, such as liver, brain, and kidney, show specific respiratory activities well above 1, and make expectedly large contributions to overall respiration. Huston and Martin (1954), working with tissue slices on supportive mats, measured the contribution of skeletal muscle as 55 per cent and of liver as 11 per cent of the total oxygen consumption.

The slight degree of inhibition of tissue slice respiration by oligomycin, not more than 30 per cent, may be due to "loose-coupling" in the mitochondria (Tobin and Slater, 1965), or perhaps to the incomplete penetration of the inhibitor. Just how large the contributions of organs to total respiration are seems to be still in question. As noted in Table 6–1, Field *et al.* were aware that their figures for liver and brain contributions were too low. Further, while the metabolic integrity of tissue slices might be more than that of homogenates, for instance, it is known that the permeability of tissue slices to oxygen and substrates is considerably limited.

PERFUSED ORGANS

Tissue slice respirometry, which usually gives a figure of 10 to 12 per cent for the contribution of liver to total body oxygen consumption, is probably a poor basis for considering the metabolic rate of the intact liver (Brauer, 1963). Thus, when liver respiration is estimated from splanchnic blood flow in anesthetized dogs, the respiratory rate is 3.4 ml O_2 per g liver per hour—a value about 70 per cent greater than that deduced from work on slices (Selkurt and Brecher, 1956). In man, the respiratory rate obtained under similar conditions is 2.2 to 2.7, and in some studies on cats, liver respiration is so high it accounts for up to 50 per cent of the total oxygen consumption.

There is argument about the applicability even of studies on perfused organs toward estimating the oxygen consumption of cells. Thus, Chance (1965) has evidence that in perfused organs the observed rate of respiration is limited by the degree to which the perfusion supplies enough O_2 to the cells, not by the avidity of the cells for O_2. The cytochrome oxidase of isolated mitochondria has an extremely high affinity for O_2, and the liver mitochondria have an affinity about the same as that of other mitochondria, not higher as the data from perfused livers might indicate. Only in suspensions of discrete cells such as yeast was enough O_2 supplied to saturate the electron-transport chain, and there it was possible to be sure that respiratory control rather than anoxia determined the rate of respiration.

Nevertheless, intact perfused organs can be used to give some information on the behavior of mitochondria. In isolated and perfused rat hearts, measurements of reflectance fluorescence show that anoxia or amytal (which inhibits respiration at the site of flavoprotein oxidation, see Fig. 3–5) keeps the intramitochondrial pyridine nucleotides in a reduced state, just as they do *in vitro*. The isolated and perfused beating rabbit heart depends upon an external supply of substrates and oxygen for the transformation of energy to support ventricular pressure-volume work. The endogenous substrates are used up in about 30 minutes, and up to 5 minutes of anoxia interferes with work done, anaerobic glycolysis failing to supply the necessary ATP. Pyruvate, acetate, and β-hydroxybutyrate are utilized by the beating heart, presumably by the myocardial mitochondria, but other substrates like succinate, malate, butyrate, and caproate are poorly utilized by the heart (probably because of permeability factors (Thorn *et al.*, 1968)).

TISSUE METABOLISM

Mammals, and other species and phylla of living organisms, have developed mechanisms for depositing and storing metabolic fuels during periods of availability, so that they can mobilize these fuels at times of increased need. The storage is effected by building up polymers or complexes of fatty acids, hexoses, and amino acids; the lipids are by far the most important form of storage of fuel, and carbohydrates are less important except for the nervous system. The stages of metabolism through which the complex compounds are utilized are discussed on page 9. The body can be divided into four main metabolic compartments that have rather distinct patterns: adipose tissue, nervous tissue, muscle, and liver.

Muscle tissue can utilize several different fuels. Glucose is glycolyzed, and the product, pyruvate, enters the mitochondrial Krebs cycle. Fatty acids and the products of their incomplete combustion, the ketone bodies acetoacetate and β-hydroxybutyrate, are oxidized by muscle mitochondria. By mechanisms not yet clarified, muscle selectively metabolizes ketone bodies, then fatty acids, and, as a last resort, glucose. If ketone bodies are unavailable and the free fatty acid concentration is low, as after carbohydrate or insulin administration, muscle will then readily metabolize glucose. Conversely, if ketone or fatty acid concentrations are high, glucose metabolism is minimized even in the presence of insulin, and the glucose which is taken into the tissue is converted to and stored as glycogen.

Liver has more complicated metabolic patterns and requirements than the other tissues and serves as a major store house of glycogen. Liver not only can oxidize glucose, lipids, and ketone bodies as an energy source but also amino acids. It also supplies substrates to the peripheral tissues. During fasting, the liver provides glucose, via the processes of glycogenolysis and gluconeogenesis, for the brain and a few other tissues which require glucose for survival. Glucose is metabolically "activated," through a phosphorylation that involves ATP, to glucose-6-phosphate in the liver, and the pattern of intra- and extrahepatic glucose metabolism is directed by two enzymes unique to liver—glucokinase and glucose-6-phosphatase. Glucokinase activity is increased by carbohydrate feeding or insulin, and glucose-6-phosphatase activity is increased by fasting, insulin deficiency, or adrenal steroids. Amino acids are the principal hepatic fuel in fed subjects, and after deamination (which supplies the urea cycle with NH_3), the carbon moieties from amino acids are oxidized via the citric acid cycle; ketone production essentially ceases. In fasted subjects, the amino acid carbon is diverted into glucose synthesis, while the oxidation of fatty acids predominates, and ketone bodies are produced; the liver citric acid cycle is inoperative.

Adipose tissue contains between 60 and 90 per cent triglycerides, which represent 6 to 8 kcal per gram of tissue. A normal 70-kg male may have 10 kg of adipose tissue and thereby a potential fuel reserve of approximately 75,000 kcal, which theoretically could support life for well over two months. In obesity, 100 kg of adipose tissue may be present—an entire year's supply! Glucose is accumulated by adipose tissue (under the influence of insulin) and by lipogenic processes it is converted to a more efficient form of energy-storing substrates, i.e., lipids; some glucose is incorporated in adipose tissue glycogen as well. Energy transformations in adipose tissue *in vivo* are as yet imperfectly understood, in part because of the known uncoupling actions of free fatty acids. One of the functions of adipose tissue involves the maintenance of body heat. Apparently adipose tissue oxidations provide heat, in addition to the known insulating properties of body fat that retain heat. Some investigators liken the *panniculus adiposus* to an electric blanket.

Some newborn mammals, hibernating mammals, and mammals which have adapted to prolonged cold exposure contain large amounts of brown adipose tissue. That tissue produces large amounts of heat, and is brown partly from the many mitochondria in its cells, with their high cytochrome content. Mitochondria isolated from brown fat by the usual methods have low P:O ratios, and it is possible that they operate in an uncoupled state *in situ* and *in vivo* to produce heat almost exclusively, rather than ATP (Smith *et al.*, 1966; Lindberg *et al.*, 1967; Thomson *et al.*, 1968). This point is under intensive investigation and is controversial.

Guillory and Racker (1968) and Aldridge and Street (1968) have shown that isolated brown fat mitochondria have normal P:O ratios when assayed in the presence of defatted bovine serum albumin. The former workers suggest the possibility that these mitochondria are uncoupled *in situ* through the presence of free fatty acids, and that hormonally induced lipolysis (norepinephrine) stimulates heat production. On the other hand, Aldridge and Street propose that the isolation of these mitochondria has damaged them and that albumin restores function to their original coupled state.* Evidence that oxidative phosphorylation in mitochondria from brown fat is fully coupled, as shown by the ability of uncoupling agents to stimulate respiration and the capacity of the cells to accumulate Ca^{2+} ions (an energy-linked process), is presented by Horwitz, Herd, and Smith (1968). When an uncoupling agent is added to these mitochondria in concentrations that stimulate respiration, the further addition of norepinephrine accelerates respiration even more. These workers conclude that in this sytem norepinephrine must act by its increasing the supply of mitochondrial substrate (the fatty acids evolved via the lipolytic effect), rather than through an uncoupling action of the free fatty acids. It is not clear, however, that because insufficient substrate limits the rapid respiration of uncoupled mitochondria, it also limits the slower respiration of coupled mitochondria.

Mitochondria prepared from the brown fat of newborn guinea pigs are partly uncoupled, having P:O ratios about half of the normal levels, but those prepared from the fat of weaned guinea pigs have higher P:O ratios progressively with weaning; all these mitochondria have little if any respiratory control, as in loose-coupling (Christiansen *et al.*, 1969). Mitochondria obtained from the brown fat of rats fail to recouple after the neonatal period. The temporal sequence in the guinea pig mitochondria suggests that the relative uncoupling does reflect the situation *in vivo*. Factors in tissue metabolism that relate to the regulation of temperature are discussed in Chapter 15.

General References: The fuel choices for the different tissue compartments are discussed in greater detail by Masoro (1968) and Hoch and Cahill (1966). Brown fat metabolism and heat production are reviewed by Smith and Horwitz (1969).

---

* These findings are analogous to those seen after injecting hypothyroid rats with thyroxine; there, the depression of respiratory control is also reversed by albumin, but in that case it is believed that the albumin removes most of the hormone from the mitochondria (see p. 91).

Part II

Physiologic Considerations: Energy Utilization

7

METABOLIC RATES

The whole animal may now be considered as the thermodynamic system we choose to measure. As in the less complex systems, the First Law must also hold. Any changes in the energy content of the whole animal must be balanced by losses or gains in the energy content of the surrounding systems. The energy balance of the whole animal can be studied by measuring the input of energy in the form of the chemical energy of foods (substrates), and the amounts of oxygen consumed in the processes of energy transformation. This energy input can be compared with the whole animal's output of energy, in the form of heat, work, and the unchanged chemical energy of the ingested or partly digested foods as they appear in the excreta (Rubner, 1894).

Heat balances are measured by calorimetry. Calorimetric measurements may be direct or indirect. In direct calorimetry, heat production is measured. In indirect calorimetry, the O_2 consumed and CO_2 produced are measured, and the chemical processes that produce heat in animals are quantitated through the use of validated assumptions.

Direct calorimetry employs temperature measurements in a closed system. Usually a weighed amount of water is insulated, and in it is immersed the heat-producing system. The rise in temperature of the water after the heat has been produced can be measured to within 0.01°C or less by thermocouples, and after corrections converted to calories* or kilocalories. The bomb calorimeter is used to measure the heat of combustion of fuels and foods, and the heat is liberated during ignition and combustion inside the sealed chamber.

Similar (without the ignition, of course) calorimetric methods are used to measure the heat produced by live animals, and this area is a relatively current one in quantitative biology. Crawford in 1778 measured animal heat production by the rise in temperature of a water jacket, and Lavoisier and Laplace in 1780 by the amount of water melted out of a block of ice surrounding the animal, using the latent heat of melting ice (80 kcal/kg of water) for calculations. At the turn of the century, more sophisticated respiration calorimeter chambers and heat-transfer and heat-insulating devices were used to validate the results obtained by indirect calorimetry. Direct calorimetry has regained a certain vogue in recent years with the appearance of a new environmental problem—living in a completely sealed system of zero gravity in space capsules. Under these circumstances it becomes very important to consider energy balances, especially over the long periods that are now feasible.

Indirect calorimetry is based on observations that biologic oxidations normally proceed via well-defined chemical pathways, and that the amount of CO_2 produced or O_2 consumed has a definite relationship to the amount of heat produced. As shown in Table 1–1, different substrates are oxidized to release different amounts of heat. Thus the amount of heat generated will depend on how much fat, carbohydrate, and protein is oxidized. The relative proportions may be deduced by measuring the O_2 consumed and the CO_2 produced, because each class of substrates is oxidized with a slightly different ratio of (CO_2 produced) per (O_2 consumed). This ratio is called the respiratory quotient or RQ. Thus:

Table 7–1

	RQ
Fat	0.7
Protein	0.8
Carbohydrate	1.0

Depending upon the RQ, one may calculate (Kleiber, 1961) that 4.7 to 5.1 kcal per liter of O_2 consumed has been liberated through oxidation. The difference in heat liberation between the various classes of substrates is thus relatively small and may be within the limits of error in measuring O_2

---

* One calorie is defined as the amount of energy required to increase the temperature of 1 g of water by 1 degree C.

consumption. In practice, advantage is taken of the fact that fasting individuals oxidize mainly fatty acids, and the value of 4.7 kcal per liter of O_2 is used. For moderate exercise in man, the value of 5.0 kcal per liter of O_2 is used.

The advantage of indirect calorimetry is that it is technically less demanding to measure oxygen consumption rather than heat production. The amount of oxygen in a system can be measured in various ways. If the closed system is filled with O_2 and the water vapor and CO_2 produced by the animal are removed by absorption on specific surfaces, the change in the gas volume remaining can be continuously measured through its pressure; or more specific methods for oxygen analysis may be used, such as chemical absorptions, infrared absorption, gas chromatography, thermal conductivity, polarography, or magnetic susceptibility. These methods all measure oxygen continuously and have the advantage over direct calorimetry that the subject need not be in the same closed system as the O_2-measuring device. A simple device for measuring the oxygen consumption of small animals like rats is shown in Figure 7–1. Other, even simpler, apparatus has been used for smaller animals, e.g., mice can be placed in a Warburg flask filled with

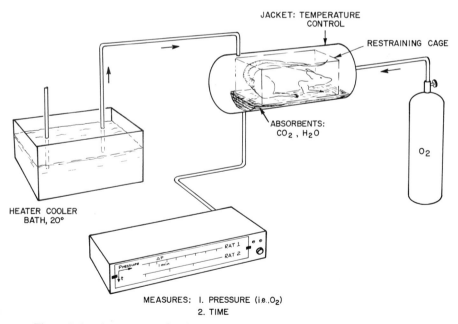

Figure 7–1. An apparatus for the measurement of the metabolic rate of small animals, weighing between about 70 g and 300 g. The animal, in a plastic restraining cage, is placed in a glass-jacketed chamber. The chamber is flushed with oxygen and contains absorbents to take up CO_2 and H_2O as they are evolved; the jacket is connected to a temperature-controlled water bath. Attached to the chamber is a volume meter (Model 160, Med-Science Electronics, St. Louis, Mo.) that measures the decrease in volume as a function of time.

O_2, and their use of O_2 can be measured manometrically. Another simple device, that can be made in the home, is described by Lauber (1969).

BASAL METABOLIC RATE

The rate of oxygen consumption or energy expenditure of an organism is expressed per mass of the organism under defined conditions to permit comparisons. For human subjects, the basal metabolic rate (BMR) is defined as the rate of O_2 consumption while the subject rests quietly, after a sleep of at least 8 hours; the last meal was at least 12 hours previously; no exercise has been performed for at least 30 minutes; and the environmental temperature is between 62°F and 87°F. For this purpose, the mass of the subject is calculated from a relationship between surface area, weight, and height, as in the DuBois chart:

Equation 7–1 $$SA = 71.84(W)^{0.425}(H)^{0.725}$$

This relationship is derived from the observation that animals of different sizes consume oxygen in proportion not to their weight but to a power function of their weight:

Equation 7–2 $$BMR \propto K(W)^n, \quad \text{where } n < 1.0$$

There has been much discussion about the best value for n, and it is between 0.67 and 0.75 (Kleiber, 1961; see Fig. 7–2). The value of K varies

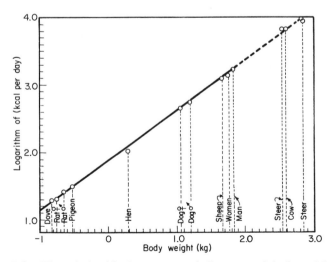

Figure 7–2. The relationship between metabolic rate and body weight in various species. The diameter of the circles represents a ± 10 per cent deviation. (From Kleiber, 1961; by permission of the author and John Wiley & Sons, Inc., New York.)

with the species. These calculations produce a number that is equated with the "surface area" of the subject. The "surface area" of a mammal is linearly related to the O_2 consumption, probably because bodies of the same shape but different size have surface areas in proportion to the two-third power of their volumes, and if their densities are the same, to the two-third power of their weights. The surface area rule is only a convenient way of expressing that exponential relationship and is not meant to imply that mammals consume oxygen by a mechanism involving their pelts. Actually, to measure the specific respiratory activity of an organism, we should know the mass of its respiratory apparatus, but that value is difficult to determine. Failing that, the lean body mass would seem a better baseline than the total body weight, which includes the variable mass of inert nonrespiring depots of fat.

There are more intrinsic factors that may vary and make comparisons difficult, to complicate determining whether a BMR is normal or not. These include age, sex, state of nutrition, athletic training, menstruation, state of relaxation, and emotional state. Further, such extrinsic factors as altitude, climate, and the time of year affect the BMR. With all this, young men and women consume about 40 kcal per hour per square meter of body surface when at rest under the conditions specified above. The basal metabolic expenditure of energy is remarkably constant. The appetite mechanism in a normal individual appears to be exquisitely set so as to maintain the body weight and to be able to compensate for the increased fuel consumption during muscular work. Thus, the weight normally varies little from year to year. For example, if an additional 100 kcal, the equivalent of one slice of bread, were to be stored daily as adipose tissue, at the end of the year there would have been over a 10-lb weight gain. The sensitivity of the appetite mechanism and of the mechanisms that expend calories beyond those required to maintain basal metabolism and the equality between these two processes in the normal individual are overtly disturbed in states of abnormal energy storage such as anorexia nervosa and obesity, but the basal metabolic rates appear to be relatively normal. Hervey (1969) hypothesizes that there is a hormonal (steroidal) effect on a central control point that balances food intake, body composition, and energy utilization.

Sleeping humans consume oxygen at rates that are 80 to 87 per cent of the rate at basal conditions. The sleeping metabolic rate thus may be considered a minimum under normal, noninhibited metabolic circumstances (if, indeed, one does not believe that sleep is caused by a "metabolic inhibitor"). It is of interest that the metabolic rate during anesthesia with various agents is in the range of 78 to 88 per cent of basal. Either all the anesthetic agents used—and they are a varied lot—actively depress oxygen consumption to the same degree, and to the same degree as the "sleep inhibitor," or (more likely) the muscle relaxation and the decrease in cardiac and respiratory work, and perhaps in brain activity in the unconscious state, are responsible for the decreased metabolic rate.

What cellular processes does the rate of oxygen consumption of the intact organism at rest represent? For one thing, cellular respiration proceeds via mitochondrial metabolism mainly, and more than 90 per cent of the O_2 consumed by the cell is reduced by the process of mitochondrial electron-transport.

The "resting" state of an intact organism represents the state wherein no external work is being performed. "Internal" work requires the expenditure of energy to support muscle contraction for respiration and the heart beat, muscle tone, and the maintenance of the continuous synthetic processes in the viscera (especially the liver), and goes on even at rest, of course. Such processes, together with the energy-requiring mechanisms that maintain the internal cellular ion-gradients, account for the energy expended by the resting organism. Mitochondrial pumping of ions, which apparently can use nonphosphorylated energy-rich bonds, proceeds in the resting state (State 4)* and so can be maintained with minimal use of oxygen. Muscle contraction uses the terminal $\sim$P-bond of ATP, generates ADP, and thereby accelerates mitochondrial and cellular respiration markedly (State 3).† The myocardium does not rest and its mitochondria are almost all in State 3. In other tissues at rest, the rate of respiration is much closer to State 4. Resting mammalian skeletal muscle has a basal rate of energy utilization of 2.5 mcal per g per minute, as measured by its utilization of phosphocreatine under conditions in which no respiration or glycolysis can occur (Schottelius and Schottelius, 1968).

THE EFFECTS OF EXTERNAL DEMANDS

Specific Dynamic Action

The ingestion of food increases the metabolic rate. Rubner originally called this the "specifisch-dynamische Wirkung," and it is usually called the *specific dynamic action* (SDA), or as Kleiber (1961) prefers to translate, the specific dynamic *effect*. The SDA depends upon the amount and type of nutrient eaten, but not on the size of the animal. Carbohydrates raise the metabolic rate above the BMR level about 4 per cent (and up to 30 per cent) for 2 to 5 hours after ingestion; fats about 4 per cent (and up to 15 per cent) for 7 to 9 hours; and proteins show the major SDA to be +30 per cent (to 70 per cent) for as long as 12 hours. It is for this reason that the BMR is defined as the metabolic rate at least 12 hours after eating.

The cause of the SDA has been a matter of dispute for decades. It was thought that the energy demands of the processes of digestion and absorption,

---

* State 4 is defined as the metabolic state in which mitochondria respire when O_2 and substrate are present in excess, but ADP + P_i are absent, resulting in a low rate of respiration.

† In State 3, respiration is rapid because of the presence of ADP + P_i.

together with the enhanced intestinal muscle contractions, accounted for the rise in O_2 consumption. However, the intravenous injection of amino acids raises the metabolic rate almost as much as the oral administration of amino acids, so intestinal mechanical and chemical work are not necessary for the SDA of proteins. There is still a question as to what portion of the metabolic pathways of amino acid utilization accounts for the SDA, i.e., catabolic or anabolic. As the increase in urinary excretion of nitrogen after amino acid intake appears to be correlated with the increase in metabolic rate, so deamination and the formation of urea are thought to be a basis for SDA. Just how these processes raise the mitochondrial metabolism of O_2 is not clear.

Liver and kidney contain an L-amino acid oxidase that has flavin mononucleotide as a prosthetic group and catalyzes an oxidative deamination that consumes O_2 in the reoxidation of the reduced flavin; H_2O_2 is also produced and is decomposed by the catalase reaction. The reactions are:

Equation 7–4

$$RCHNH_2 \cdot COOH + FMN + H_2O \rightleftharpoons RCO \cdot COOH + NH_3$$
$$+ FMNH_2$$

$$FMNH_2 + O_2 \longrightarrow FMN + H_2O_2$$

$$2 H_2O_2 \xrightarrow[\text{catalase}]{} 2 H_2O + O_2$$

It will be seen that a half-mole of oxygen is consumed for every mole of L-amino acid oxidized; however, this enzyme has a very low turnover number and does not oxidize dicarboxylic and dibasic amino acids, or serine and threonine. The L-amino acid oxidase probably accounts for none of the SDA. Liver mitochondria do oxidize glutamate through an initial transamination to α-ketoglutarate, using the oxaloacetate transaminase to form aspartate:

Equation 7–5

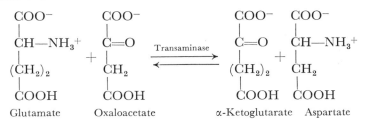

rather than using the glutamic dehydrogenase (which would perform an oxidative deamination). Glutamate derived from protein hydrolysis would only supply additional substrate for mitochondria, and substrate is not the limiting factor in mitochondrial respiration. Urea formation uses ATP; the formation of carbamyl phosphate in mammalian liver produces ADP, and the ATP used to form the acyl guanidinium linkage between citrulline and aspartate is split to form pyrophosphate and AMP. The ADP produced might stimulate mitochondrial respiration and result in calorigenesis.

Equation 7–6a $\quad CO_2 + NH_3 + 2\ ATP + H_2O \xrightarrow[\text{synthetase}]{}$

$$H_2N-CO-OP + 2\ ADP + P_i + H^+$$
$$\text{carbamyl-P}$$

Equation 7–6b

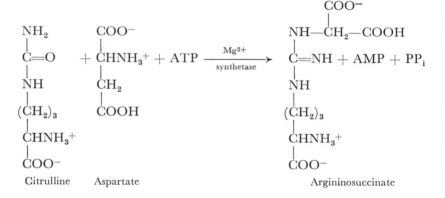

| Citrulline | Aspartate | | Argininosuccinate |

The anabolic processes of amino acid incorporation into peptides and proteins certainly do use energy-rich phosphate bonds. The activation of the amino acids involves a pyrophosphorolysis of ATP, and ribosomal translation of the activated amino acyl-tRNA complexes involves the use of the high-energy terminal bonds of GTP (which are ultimately derived from ATP by a kinase enzyme). The rephosphorylation of AMP to ADP and eventually to ATP by mitochondrial oxidative phosphorylation would seem a possible source of the observed calorigenesis in SDA; however, whether the amount of protein synthesis and the amount of ADP generated are enough to account quantitatively for the extra uptake of oxygen has not yet been measured.

The role of protein synthesis in raising the metabolic rate in intact animals has been invoked by some investigators as the source of the calorigenic action of the thyroid hormones as discussed in Chapter 8. Injected thyroid hormone increases the rate of amino acid incorporation into proteins in as early as 2 hours (Sokoloff *et al.*, 1968). If protein synthesis is the basis for SDA, the hypotheses of Sokoloff and of Tata may be reduced to a hormonal stimulation of SDA. The thyroid hormone is still calorigenic in starving animals, where no protein synthesis occurs, so an "SDA-like" hormonal effect may contribute only a part of the calorigenic effect of the hormone.

An increased formation of catecholamines, which increases the metabolic rate, has also been proposed as a basis for the SDA. Catecholamines are synthesized metabolically from the amino acid tyrosine, and the ingestion of tyrosine involves an excess production of such amines as epinephrine (Abelin and Goldstein, 1954, 1955).

Work

In active subjects, the amount of food ingested is proportional to the amount of energy expended, and so the body weight remains constant. In sedentary or immobilized subjects, the same food intake will lead to an increase in body weight, usually as fat. Conversely, as activity increases, an increased caloric intake is necessary to maintain body weight, until the degree of activity increases to the point of exhaustion, in which case even a very large food intake does not prevent weight loss.

Work, or muscular contraction, raises the metabolic rate above the basal level, and the amount of work performed is roughly proportional to the rise in metabolic rate. Thus, moderate work performed by a 70-kg man may expend up to 1900 kcal per 8-hour day and raise oxygen consumption up to three times the basal rate. Such work loads are well-tolerated. Hard work raises the metabolic rate from four to eight times the basal rate and is a strenuous day-to-day load. Maximal work raises the metabolic rate above eight times basal, and, for short periods, the rate may be very high indeed. Therefore, a 154-lb man may expend 70 kcal per hour while asleep, about 600 kcal per hour while walking up stairs at 2 mph, and 9500 kcal per hour while running on the level at 18.9 mph (for very short periods). When a man weighing 66 kg runs on the level at 6 mph, he expends 11.6 kcal per minute (2.4 liters O_2 per minute; 8350 kcal per day). If he runs at 8.7 mph (+45 per cent), the energy cost mounts to 15 kcal per minute (+29 per cent). If his body weight is increased by 5 kg (+7.5 per cent), running at 6 mph expends 12.5 kcal per minute, and at 8.7 mph, 16.1 kcal per minute (7 to 8 per cent). Energy expenditure thus increases proportionately to the speed of running. To consume excess body weight in an exercise program it would seem that jogging while carrying weights is more effective than just running slightly faster.

Exercise increases and inactivity decreases the capacity of mammals to perform mechanical work. Part of these effects arises from changes in the amount of contractile muscle fibers, e.g., exercise somehow causes an increase in the diameter of muscle fibers and the synthesis of more actomyosin. Another part apparently arises from changes in the composition of the *cristae* of skeletal muscle mitochondria. In exercising rats, as compared with sedentary controls, the mitochondrial protein per gram of gastrocnemius muscle is increased about 60 per cent, and the activity of cytochrome oxidase and cytochome c is increased about 100 per cent (Holloszy and Oscai, 1969). The activity of mitochondrial α-glycerophosphate dehydrogenase per gram of muscle is not changed, indicating that exercise induces the synthesis of specific mitochondrial electron-carriers rather than simply an increase in mitochondrial size or number. The net result is an increased capacity to transform energy to ATP, perhaps from fats more than from carbohydrates.

A well-trained human athlete can expend energy at 15 to 20 times his basal rate for only a few minutes. In contrast, some birds can fly at 20 to 30 mph continuously for up to 2000 miles without feeding, using from 0.02 to 0.05 per cent of their body weight in fat for each mile traveled to expend energy at about 15 times their basal rates (Tucker, 1969). Mammalian energy cost of body-transport can be compared with that of other creatures and of machines in units of calories required to transport a gram of body weight one kilometer (Fig. 7–3). The cost generally goes down as body weight goes up. Walking and running mammals expend 10 to 15 times more energy per distance traveled than a bird of the same size, but less than some machines.

The resting rate of metabolism can be raised up to 130-fold under conditions of maximal exercise. That degree of increase in oxygen consumption is an order of magnitude greater than can be accounted for by the stimulation of mitochondrial respiration from State 4 to State 3 when ADP is added. Yet even at the highest rates of O_2 consumption, mitochondrial electron-transport must be involved. One answer to this apparent paradox might be that in the resting state muscle mitochondria respire *in vivo* even more slowly than they do *in vitro*, and that a greater proportion of the total oxygen consumption of the organism is accounted for by visceral organs like the liver, heart, and brain than appears from summations of tissue respirations *in vitro*. Then the rise in total O_2 consumption when muscle mitochondria are exposed to ADP during muscle contraction would make a large

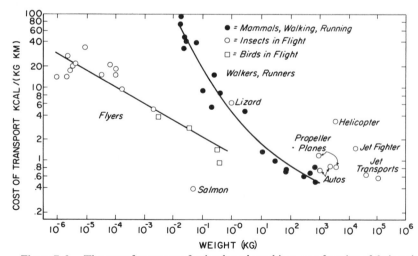

Figure 7–3. The cost of transport of animals and machines, as a function of their weight. Each point represents, wherever possible, animals or machines flying, walking, running, or cruising at the speed at which the cost of transport is least. (Courtesy of Dr. V. A. Tucker.)

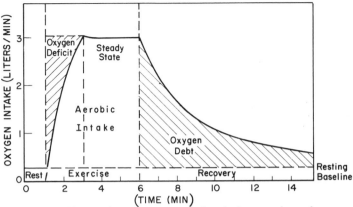

Figure 7-4. Changes in oxygen consumption during exercise and recovery.

contribution to the metabolic rate, although muscle mitochondria make a small contribution during rest.

At the moderate rates of energy expenditure during work after a steady state has been reached, the amount of oxygen consumed is about linear with the work performed. At higher rates, the oxygen consumption levels off, presumably reflecting the maximal capacity of mitochondrial electron-transport when stimulated by ADP, i.e., State 3. If the work done increases above that level, the extra energy must be supplied by mechanisms that do not use oxygen, and then anaerobic glycolysis is markedly accelerated. Rapid anaerobic glycolysis produces lactic acid faster than lactic acid can be disposed of, and large amounts may accumulate. When it is recalled that glycolysis is energetically about $\frac{1}{18}$th as efficient as oxidation, the accumulation of large amounts of lactic acid becomes understandable.

When the work performed is mild or moderate, a steady state of O_2 consumption, CO_2 elimination, body temperature, and blood lactate concentration is reached within 1 or 2 minutes (Fig. 7-4). Even when the amount of work done starts and remains at a constant level, the rate of O_2 consumption does not rise to its steady rate immediately but rises rapidly over the first 2 minutes or so. The performance of work during this period of disparity between energy supply and demand—the oxygen deficit—is supported anaerobically through the breakdown of ATP. The amount of ATP stored in mammalian skeletal muscle is about 4 m moles per kg wet weight (Piiper *et al.*, 1968), enough to support only one to two twitches of the fibers; however, mammalian skeletal muscles contain about three to six times as much phosphocreatine. The phosphate of phosphocreatine is a high-energy group and can be enzymatically transferred to ADP without evoking an additional source of energy, i.e., anaerobically:

Equation 7-7 $Cr \sim P + ADP \underset{\longleftarrow}{\overset{\text{creatine phosphokinase}}{\longrightarrow}} Cr + ATP$

The free creatine generated can be phosphorylated back to creatine phosphate by the reverse reaction; the resynthesis may occur during the steady state after 2 minutes (Fig. 7–4). Oxidative phosphorylation can supply enough ATP to support both muscle contraction and the reverse reaction in Equation 7–7, only if the work load is small.

If the work load is great so that oxidative phosphorylation cannot support muscle contraction, anaerobic glycolysis is evoked to supply ATP, and thereafter both lactic acid and creatine will accumulate. When work stops, oxygen consumption stays above the resting rates for hours; this excess O_2 consumption is the oxygen debt. The amount of the oxygen debt incurred varies with the athletic training of a subject. Untrained persons have a maximum O_2 consumption rate of 2 to 3 liters per minute. Trained persons may consume up to 5 liters per minute and so have a greater aerobic work capacity. The tolerance to the oxygen debt also increases with training, up to 15 liters in an athlete. In supramaximal exercise, first $\sim$P-bonds are transferred anaerobically until no more remain, then glycogen is split to lactic acid anaerobically to supply $\sim$P. In strenuous intermittent exercise, no lactic acid is formed if the oxygen debt contracted during work can be met completely by the oxidative resynthesis of $\sim$P bonds via oxygen consumption during the rest periods. In this way, a greater total amount of work can be done by strenuous intermittent exercise than by exercise continued until exhaustion (Margaria *et al.*, 1969).

The components of the early oxygen debt seem to be the resynthesis of ATP and phosphocreatine (rapid) and the removal of lactic acid. Glycogen is resynthesized in muscle from the accumulated lactic acid, using $\sim$P-bonds. Even when these components are taken into account, the persistence of a raised metabolic rate for hours is still not explained.

General References: Early works on metabolic rates are DuBois (1936) and Brody (1945); Kleiber's (1961) text should be read. Margaria (1966) and Morehouse and Miller (1967) review the biochemistry and the effects of exercise.

CHAPTER

8

THYROID
HORMONES

The thyroid hormones control energy transformations. That has been clear since 1895, when Magnus-Levy showed that the thyroid hormones controlled the rate of oxygen consumption in mammals. This understanding was based upon observations of spontaneous disease states, and of the effects of administration of thyroid hormones or ablation of the thyroid gland. Excess of hormone raises the rate of oxygen consumption above normal, deficiency of hormone depresses oxygen consumption below normal, and administration of small amounts of hormone to deficient subjects restores oxygen consumption to normal. From such observations it has been concluded that the function of the thyroid hormones is to act as a pacesetter or thermostat that controls the rate of cellular oxidations and thereby energy transformations.

The thyroid hormones are iodinated amino acids. The two most important ones are L-thyroxine (L-3,5,3′,5′-tetraiodothyronine, L-T$_4$) and L-3,5,3′-triiodothyronine (L-T$_3$) (Fig. 8-1).

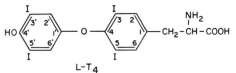

Figure 8-1. The structure of L-thyroxine (L-T$_4$) and L-triiodothyronine (L-T$_3$).

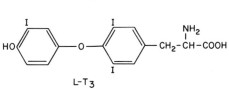

A great deal is known about the synthesis and the metabolism of the thyroid hormones, mainly by taking advantage of the fact that iodine is a unique component and can be measured specifically and sensitively by chemical methods or by substituting for the natural ^{127}I the radioactive isotope ^{131}I. Somatic cells, and the thyroid gland, contain iodinated thyronines other than $L\text{-}T_4$ or $L\text{-}T_3$, as well as iodinated tyrosines and free iodide, but these components all seem to be precursors or metabolites of $L\text{-}T_4$ and $L\text{-}T_3$. There is still a question as to whether some of the other iodinated thyronines are even more active in the cell than $L\text{-}T_4$ or $L\text{-}T_3$, and are the actual "active" forms of the hormone. The acetic acid derivatives Tetrac and Triac are particularly active. But by far the greatest amount of iodine-containing material in cells is $L\text{-}T_4$ and $L\text{-}T_3$, and they are usually referred to as the "thyroid hormones."

The thyroid hormones are synthesized only in the thyroid gland. The rate of synthesis depends upon (1) the amount of iodide supplied to the thyroid gland (iodine is a trace element in mammalian bodies, but even the minute amounts of iodine needed must be supplied in the food or water), and (2) the presence of a specific "thyroid-stimulating hormone" (TSH), a polypeptide hormone secreted by the anterior pituitary gland. A feedback control system acts so that increased amounts of free-circulating thyroid hormones depress TSH production, and there is a second controlling system over the anterior pituitary whereby the hypothalamus secretes a "thyrotropin-releasing factor" (TRF) that acts on the pituitary to stimulate the release and perhaps the synthesis of TSH. Porcine and ovine TRF have recently been identified as a simple tripeptide, consisting of Glu-His-Pro. The N-terminal Glu is in a ring structure (2-pyrrolidone-5-carboxylyl or *pyro*glutamyl) and the Pro is present as the amide (Fig. 8–2). This compound has been synthesized, and 10 $\mu\mu g$ of the synthetic or the natural hormone is active *in vitro*, as reported from Guillemin's laboratory (Burgus *et al.*, 1969).

The details of these systems are discussed in a number of endocrinology texts. For the purpose of understanding the regulation of energy metabolism it is important to know what physiologic stimuli turn thyroid hormone production on and off, but those are not yet well understood. It appears that cold environments stimulate, and hot environments depress thyroid hormone synthesis, both via hypothalamic centers. This seems teleologically

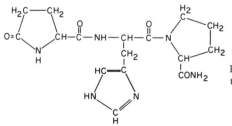

Figure 8–2. The structure of the hypothalamic thyrotropin-releasing factor (TRF).

L-(pyro) GLU-L-HIS-L-PRO (NH₂)

reasonable, low temperatures producing a demand for more heat, and high temperatures for less heat. But the thyroid hormone responses to continued temperature extremes are only temporary; for instance, when animals are kept at temperatures near freezing, thyroid hormone production rises above normal only for a period of a few weeks, then falls to normal again. Very recent studies indicate that the amounts of thyroid hormone in the mitochondria of some tissues may vary sharply and rapidly (within 2 hours) in response to moderate temperature changes, suggesting that the notion of thyroid hormones as only long-range pacesetters needs reconsideration. Possible reasons for this will become apparent when we discuss how the thyroid hormones act on cells. There are undoubtedly other stimuli than temperature that also control thyroid hormone synthesis, such as emotional stresses. There is the complex problem of how thyroid hormone secretion depends upon and affects other hormones.

LATENT PERIOD FOR CALORIGENESIS

The thyroid hormones alter the basal metabolic rate after a latent period of days. The effects of administered thyroid hormone upon the BMR can be seen most clearly in completely hypothyroid subjects, because there the variability of endogenous hormone production is absent. On the other hand, athyroid subjects are, for some reason not yet understood, extremely sensitive to the hormone, and doses that have no effect on the metabolic rate of a euthyroid subject are calorigenic; whether the latent period is also different is not sure, but, in general, the changes seem to appear faster in hypothyroidism. In a severely hypothyroid patient, with a BMR of -45 per cent, a single intravenous dose of 10 mg of thyroxine increased the metabolic rate to about -6 per cent only after 7 days (Means and Lerman, 1935). L-T_3 acts with a shorter latent period, the first increases occurring at 6 to 12 hours, and the maximum being obtained at about 36 hours. The difference between the latent period of L-T_4 and L-T_3 is marked in humans, because L-T_3 is bound less firmly than L-T_4 to the serum proteins, which allows a higher serum concentration of free L-T_3, the active form. There is much less difference between the latent period of the two in rats, where the serum proteins bind L-T_4 more weakly.

Similarly, there is a slow "latent period of deactivation" during which calorigenesis returns to the initial level. It took 30 days for the BMR to fall back to -30 per cent in Means' and Lerman's athyroid patient.

The existence of a latent period of days for the calorigenic effects of thyroid hormones is in striking contrast to the behavior of other calorigenic agents, like 2,4-dinitrophenol and the catecholamines. The latter raise the BMR in 15 to 30 minutes, and, depending on the dose, maintain the raised BMR for an hour or so, after which there is a rapid fall to normal levels at about 2 hours. Since it is clear that DNP raises the BMR through a direct action on mitochondrial respiratory control, the existence of a long latent period for the action of the thyroid hormones has been a strong argument against accepting a mechanism of calorigenic hormone action that appears similar to that of DNP.

Various explanations have been advanced for the long latent period of the calorigenic action of the thyroid hormones. It has been thought that the latent period might reflect the time required for the distribution of the hormone to intracellular compartments and to the target sites. If this were so, it would allow for a calorigenic action that was direct, like that of DNP. But administration of thyroid hormone raises tissue hormone contents extremely rapidly, almost as quickly as one can sacrifice the animal and isolate tissues or subcellular fractions; and the half-life of the hormone in tissue fractions or blood is of the order of minutes or of 12 to 24 hours. Another explanation that permits direct hormonal calorigenesis has been that the delay must be largely due to the time required for the transformation of L-T_4 and L-T_3 into an active form in the cells, like Tetrac or Triac. But there is no evidence for such active forms existing to any major degree in tissues at the peak of calorigenesis after hormone injection; tissues convert the hormone to these and other congeners to a minor degree and rather rapidly, and so a long delay would not be accounted for.

The most plausible explanation for the long latent period has been that the effects on intracellular enzyme components are slow, and that a train of events is set off by hormone injection that takes days to result in the synthesis of enough oxidative enzymes to raise the BMR. This explanation does not allow for a direct hormonal activation of oxidative enzymes, but removes the direct action to a different system, e.g., transcription and translation, and to a much earlier time after hormone injection.

Any proposed mechanism for the action of the thyroid hormones must account for the latent period; however, one must be careful to specify *the system that is being measured.* Thus, it seems reasonable now that the major, long-lasting rise in the BMR—days after hormone injection—is due to the slow and delayed synthesis of oxidative enzymes and mitochondrial respiratory assemblies. This does not eliminate, however, the occurrence of a rapid and transient calorigenesis that reflects a direct action of the hormones on oxidative enzyme activity; Donhoffer *et al.* (1949, 1958) and Kaciuba-Uscilko *et al.* (1970) have demonstrated this under special circumstances. In a sense, if one accepts that hormones act by interacting at a target site, there is essentially no latent period if the effect of the hormone is measured at the proper site.

HORMONAL ACTIONS AND EFFECTS

The thyroid hormones control not only oxygen consumption, but also other processes in a number of organs and metabolic systems. Excess or deficiency in the amount of thyroid hormones modifies temperature regulation, growth and development, the response to other hormones, nerve function, and the metabolism of proteins, fats, carbohydrates, nucleic acids, vitamins, and inorganic anions and cations. On the other hand, thyroxine and triiodothyronine are relatively simple molecules, and their small size and limited number of reactive groups suggest either that the variety of the effects they produce are due to a few types of primary interactions at the molecular level, or that the hormones are changed in the cell to analogues, each having a different and specific physiologic effect. The preponderance of evidence at present supports the first hypothesis. In discussing how the thyroid hormones act at the molecular level, and how the physiologic and chemical changes are related to the primary event, one must differentiate between an *action* and an *effect*.

Actions may be defined as those functional or structural changes that are primary and depend upon the presence of the hormone at a site where it interacts with molecules in the cellular apparatus. Because hormones are effective in small amounts, we may assume that their primary molecular interactions are reversible, so that the hormones are not used up. Because of the disparity between the paucity of a hormone and the relatively much larger degree of change exerted by its presence, it has seemed reasonable to suppose that the primary actions of hormones are exerted at the site of some sort of amplifying system. An enzyme, with its high turnover number, or a "permeability" site in a membrane, with its control over the passage of a large number of molecules, has seemed—and still seems—a logical choice for a site of action. Hormonal mechanisms of direct action may proceed via allosteric changes in enzyme or protein conformation. Some investigators prefer to account for the multiplicity of changes induced by hormones by postulating several sites of action.

The control exerted over metabolism by direct hormonal actions is called "fine" control. Function changes rapidly (instantaneously or with a minutes-long lag period) after hormone injection *in vivo* or after addition of hormone *in vitro*.

Effects of hormones may be defined as those functional, structural, or compositional changes that are secondary and do not depend upon the presence of the hormone. Effects should not be reversed if the hormone is removed after exerting its primary action. The control exerted over metabolism in this category is called "coarse" control; it is slow, with a latent period of hours or days. The sites of hormonal effects appear earliest in the cellular process of information storage and transmission, and are finally expressed via the synthesis of proteins, and especially enzymes.

In brief, hormones can control both the activity of enzymes and the amount of enzymes.

The thyroid hormones are peculiarly suitable for the resolution of primary actions from secondary effects, because it is possible to measure the presence of thyroid hormones specifically and relatively easily. The iodine moieties of thyroxine and triiodothyronine can be used experimentally as a tracer for quantitative analysis. Methods for detecting other hormones not possessing this useful property are less specific and sensitive, or more tedious.

The thyroid hormones exert both fine and coarse control over energy transformations. The thyroid hormones act directly and rapidly or instantaneously upon mitochondria to change the specific activity of the electron-transport chain; and the thyroid hormones effect secondarily, and after hours or days, an increased rate of protein synthesis, part of which is manifested in an increased amount of mitochondrial electron-carriers. The actions on mitochondria appear to cause the effects on protein synthesis.

Actions of Thyroid Hormones

As the processes of mitochondrial oxidative phosphorylation became better understood, the laboratories of Lardy (Lardy and Feldott, 1951), Lipmann (Niemeyer *et al.*, 1951), and Martius (Martius and Hess, 1951) noted the similarities between the phenolic structures of thyroxine and 2,4-dinitrophenol, and between the physiologic manifestations of DNP poisoning (see Chapter 12) and thyrotoxicosis. In *in vivo* and *in vitro* experiments on normals rats, the thyroid hormones were shown to affect liver mitochondria like DNP did: thyroxine increased controlled State 4 respiration and uncoupled oxidative phosphorylation. Very large doses of L-thyroxine (L-T_4) were used; a total of 28 mg per 200 g rat (140 μg per g) was given subcutaneously over 4 days in Hoch and Lipmann's studies (1954). *In vitro*, concentrations of L-T_4 between 10 μM and 100 μM were necessary to stimulate respiration and to uncouple oxidative phosphorylation. Even at that time it was recognized that these catabolic and energy-wasting actions of the hormone served as a rationale only for the toxic manifestations seen in thyrotoxicosis, but not for the anabolic and energy-conserving manifestations that smaller doses of thyroid hormones evoked in hypothyroid subjects. Nor were the symptoms and signs of hypothyroidism made more understandable by the "uncoupling" hypothesis. Further, although thyroxine seemed to act on mitochondria like DNP, DNP could not always substitute for thyroxine: DNP did not cure hypothyroidism, or cause metamorphosis in the *Anura*, for instance. For such reasons, attention was directed away from the mitochondrion in the search for a mechanism of anabolic, energy-conserving, constructive thyroid hormone actions; such extramitochondrial effects, but not actions, were found, as is described in the next section.

However, not all of the actions of the thyroid hormones upon mitochondria are actually energy-wasting, just as not all of the actions of DNP are. Small doses of thyroxine injected *in vivo*, and low concentrations added *in vitro*, stimulate controlled mitochondrial respiration without appreciably lowering the efficiency of phosphorylation; the hormone can produce loose-coupling (see Chap. 5). Loose-coupling is an anabolic and energy-conserving process, and might be a basis for the constructive effects of the hormone

(Lardy, 1954). We know now that State 4 respiration can provide energy in the form of nonphosphorylated energy-rich bonds that are capable of supporting such processes as ion accumulation (Carafoli *et al.*, 1965) and perhaps protein synthesis; an accelerated State 4 respiration can supply more energy. As will be seen, the hormonal stimulation of extramitochondrial protein synthesis may be mediated by an action of the hormone on mitochondria.

Recent studies have shown that the thyroid hormones exert a direct action upon the mitochondria of rat liver that has the features of being energy-conserving, very rapid, sensitive to small doses, and dependent upon the hormone's presence. Hypothyroidism alters the mitochondrial respiratory responses to specific activators ($ADP + P_i$, and DNP) and inhibitors (oligomycin, guanidine, rotenone, antimycin *a*), and injection of the hormone restores some of the responses to normal levels in a surprisingly short time. The amount of the hormone in the mitochondria has been correlated with some of the changes in respiratory function.

Respiratory control is the regulation of mitochondrial respiration through the presence of $ADP + P_i$, by alterations in the specific activity of the enzymes of the electron-transport chain (see Chap. 5). *In vivo*, resting cells respire in the controlled state (State 4), as if insufficient $ADP + P_i$ were present. There is experimental evidence that the thyroid state controls the respiration of mitochondria in State 4, and thereby the basal metabolic rate. In hypothyroid rats, liver mitochondria respire slowly in State 4 owing to excessive respiratory control (Maley and Lardy, 1955; Fig. 8–2). The specific activity of maximal electron-transport (State 3) is not as severely limited as is controlled respiration, being at a level close to that seen in mitochondria from the livers of normal rats. The correlation between State 4 respiration and the BMR in hypothyroidism is quite good; the former is −45 per cent below the normal rate, and the BMR is −40 per cent in severe hypothyroidism (see Fig. 8–5). (The changes in the BMR in hypothyroidism, and after hormone-injection, are probably not due solely to changes in the specific activity of the oxidative enzymes, in spite of this correlation and other equally good ones seen in hyperthyroid animals. As will be discussed below, the *amount* of the oxidative enzymes and cofactors is also under thyroid influence, and the temporal correlations between the BMR and the amount of respiratory assemblies is even better than that between BMR and State 4 respiration. This is still an unsettled point.)

Figure 8–3 shows the degree of abnormality in respiratory control and controlled respiration in liver mitochondria from untreated hypothyroid rats, and the rapid restoration toward normal after injection of L-T_4. Significant alterations are seen in as early as 2 minutes, and respiratory function becomes normal in 3 hours. Bronk (1966, 1968) and Volfin and Sanadi (1968a, b) have confirmed these findings. Pretreating normal rats with L-T_4 raises liver mitochondrial respiration significantly above normal as early as 6 hours (Hoch, 1968a); again, this is a very short latent period

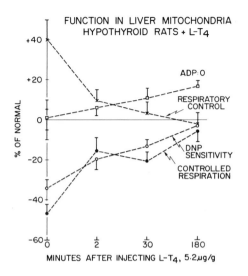

FUNCTION IN LIVER MITOCHONDRIA
HYPOTHYROID RATS + L-T4

MINUTES AFTER INJECTING L-T₄, 5·2μg/g

Figure 8–3. Oxidative phosphorylation in mitochondria obtained from the liver of hypothyroid rats before and after the intraperitoneal injection of L-thyroxine (L-T$_4$), 5.2 μg per g. Controlled respiration (State 4), respiratory control (the ratio, State 3/State 4), and the ADP:O ratio were measured polarographically with β-hydroxybutyrate as substrate. Sensitivity to dinitrophenol (DNP) is represented by the value of n, where $\log [v/(V_m - v)] = n \log (\mathrm{DNP}) - \log K$ (see Hoch, 1968b); glutamate was the substrate. Values are presented as mean per cent differences $\pm$ SE from those in mitochondria from normal rats. The abscissa shows the time between injection of the hormone and sacrifice of the animal. (From Hoch, 1967a.)

compared to the 72 hours of latent period before new mitochondria are synthesized in such animals.

Such depressions of respiratory control are reminiscent of the actions of administered uncoupling agents (see Chap. 12). They are not, however, a sign of uncoupling in the experiments discussed here. The ADP:O ratio is not decreased after the administration of L-T$_4$ (Fig. 8–3). The inhibitory action of oligomycin, which is specific for phosphorylating respiration, and is normal or even greater than normal in hypothyroid mitochondria, is not decreased by injecting the hormone under these conditions. L-T$_4$ here acts to conserve energy.

To adduce a primary action of thyroxine on mitochondria, the administered hormone should be present in mitochondria at a time consistent with the changes in mitochondrial function. The presence of the hormone has been measured (Dillon and Hoch, 1967) by assaying total iodine content, butanol-extractable iodine, and butanol-insoluble iodine in isolated liver mitochondria (Fig. 8–4). Normal liver mitochondria contain 0.43 μg of total iodine per g protein, of which a large proportion represents hormone: 52 per cent is butanol-extractable and another 10 per cent remains protein-bound. Mitochondria prepared from the livers of hypothyroid rats contain only 20 per cent of the normal amount of iodine (mitochondria from hypothyroid rats are hypothyroid). Injecting the hypothyroid rats with L-T$_4$ raises the content significantly (57 times normal) at 2 minutes and progressively for 3 hours to 130 times normal. The changes in oxidative phosphorylation are approximately proportional to the logarithm of the mitochondrial hormone content. The amount of hormone arriving in the mitochondria at 3 hours is stoichiometrically significant, being one to five molecules per respiratory assembly and up to 50 μM with reference to the water content of the cristae. Thus, the very short latent period and the presence of significant amounts of the hormone support the conclusion that

Figure 8–4. Iodine content of mitochondria from the livers of hypothyroid rats before and after the injection of L-thyroxine, as in Figure 8–3. The means $\pm$ SE of values for total iodine (TI), butanol-extractable iodine (BEI), and butanol-insoluble iodine (BII) are shown in mμg I per mg of mitochondrial protein. The ordinates at the right show the calculated amounts of thyroxine (T_4) in mitochondria, assuming $4I = 1\ T_4$. On the scale of $\mu\mu$ moles of T_4 per mg of mitochondrial protein is also shown the amount of mitochondrial cytochrome c, as measured in mitochondria from untreated hypothyroid rats by Maley (1957) (the higher value) and by Hoch (1965c) (the lower value); "Rot" indicates the amount of the rotenone-sensitive catalyst. The scale of $\mu M\ T_4$ is calculated from values for the available volume of mitochondrial water and represents an apparent intramitochondrial hormone concentration. (From Hoch, 1967a.)

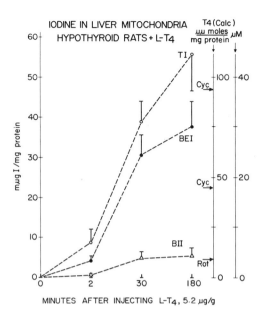

MINUTES AFTER INJECTING L-T4, 5.2 μg/g

the hormone acts directly on an enzyme complex associated with mitochondria. Further support comes from observations that the hormone is localized in the submitochondrial particles prepared from intact mitochondria by drastic sonication (Arbogast and Hoch, 1968), because these particles represent fragments of the inner membrane that contain the oxidative and phosphorylative apparatus.

The early hormone-induced changes in respiratory control are reversible. Bovine serum albumin binds thyroxine strongly, and can remove thyroxine from mitochondria, either when the hormone is added to the mitochondria or when the animal is injected with hormone. In Table 8–1 are shown the respiration and iodine content of liver mitochondria from hypothyroid rats injected with L-T_4 30 minutes before sacrifice; the same mitochondria were assayed without albumin, or in the presence of (or after extraction with) albumin (Hoch and Motta, 1968). L-T_4-injection in hypothyroid rats raises State 4 respiration and depresses respiratory control

Table 8–1 *Effects of Albumin on Liver Mitochondria*

| | HYPOTHYROID RATS | | | | NORMAL RATS | |
| | Controls | | L-T_4 Treated | | | |
	$-Alb$	$+Alb$	$-Alb$	$+Alb$	$-Alb$	$+Alb$
State 4 Respiration	3.8	3.8	5.2	3.8	5.9	5.3
Respiratory Control	4.2	4.9	3.3	5.2	3.5	4.0
Total I (μg/g)	0.35	0.38	19.4	5.3	0.71	0.40

significantly when measured in the absence of albumin. In the presence of albumin, the same mitochondria show no functional changes after L-T$_4$-injection. Albumin removes about 75 per cent of the hormone (butanol-extractable and butanol-insoluble iodine to the same extent as Total I) from the treated rats' mitochondria, but none from the controls; albumin also removes the hormone from normal mitochondria. Probably this effect of albumin on respiratory control depends not only on the removal of the hormone, but also on the removal of endogenous fatty acids or Ca^{2+} ions or other substances as well, because albumin improves respiratory control in mitochondria from untreated hypothyroid rats without removing iodine. Although the mitochondria from L-T$_4$-treated rats contain 15 times the control (hypothyroid) amount of hormone after extraction with albumin, their respiratory control is as high as or higher than the respiratory control in the untreated hypothyroid animals. The hormonally induced changes depend at least in part upon the presence of the hormone, and so may be considered the results of the action of the hormone.

These findings also explain why some workers (Tata *et al.*, 1963) have seen no changes in mitochondrial respiratory control after giving relatively small doses of thyroid hormone *in vivo*. Bovine serum albumin is frequently added to assay mixtures in measuring oxidative phosphorylation because, as shown in Table 8–1, it "improves" respiratory control. When albumin is so added in assaying the mitochondria from thyroid-treated animals, as in Tata's studies, only the changes in function that are not reversible will be observed. Under such conditions, it appears that the thyroid hormones have only slowly evolving effects on the amount of respiratory enzymes in mitochondria. Albumin, or preferably an agent more specific for thyroid hormones, may be used to differentiate the actions from the effects of the hormone.

If the thyroid hormones act directly and rapidly upon mitochondria *in vivo* and thereby raise the rate of respiration in State 4, and if mitochondria respire in State 4 *in vivo*, the metabolic rate of the animal should be increased very soon after injection of the hormone, since the hormone arrives in mitochondria almost immediately. Most investigators believe that the thyroid hormones raise the BMR only after a latent period of many hours to days, and this has been taken as a serious counter to the significance of the evidence for very rapid hormonal actions on oxygen consumption. However, some investigators report immediate calorigenic effects of thyroxine. Donhoffer *et al.* (1949, 1958) saw immediate calorigenesis in some hypothyroid, hypophysectomized rats. More recently, Kaciuba-Uscilko *et al.* (1970) find that the metabolic rate and the rectal temperature of suckled pigs aged 1 to 11 days rises about 4 hours after the subcutaneous injection of 80 mμg of thyroxine per g; a second rise, after daily injections of the hormone, occurs in 5 to 7 days. The age of the animals is crucial in demonstrating the immediate calorigenesis, perhaps because the hormone is bound to plasma proteins to a greater degree in older animals.

Figure 8–5. Effects of L-thyroxine on the calorigenic response to DNP in a hypothyroid rat. The basal metabolic rate (BMR, x, left ordinate) and the per cent calorigenic response to DNP (↑, 10 μg/g, right ordinate) are shown as a function of time. L-thyroxine was injected intraperitoneally in doses of 5.2 mμg per g, once a day; equivalent volumes of saline solution were injected initially during the control period. (From Hoch, 1965b.)

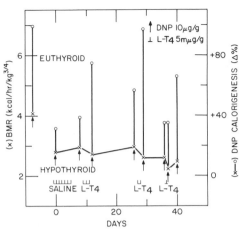

Mitochondrial respiratory responses to DNP are controlled by the thyroid state; this appears to be one case of a generality that all calorigenic actions depend upon the thyroid state (see Chaps. 12 and 14). For instance, in normal rats, a dose of 10 μg of DNP per g body weight raises the metabolic rate (by acting on mitochondria) about 60 per cent (Fig. 8–5). A hypothyroid rat (with a BMR 40 per cent below normal) responds to the same dose of DNP with a lesser degree of rise in metabolic rate, about +25 per cent. A very small dose of L-T$_4$ (5.2 mμg per g), less than that required to raise the BMR, restores DNP-induced calorigenesis to normal levels with a latent period of only 3 to 6 hours. The degree of DNP-induced calorigenesis depends reversibly upon the thyroid state: 6 to 10 days after thyroid injection has restored the normal degree of DNP-calorigenesis, a second testing with DNP produces a subnormal response again; the second response can be restored to normal by a second injection of hormone.

DNP raises the metabolic rate by acting directly upon mitochondrial respiratory control. The subnormal calorigenic responses of hypothyroid rats to DNP injected *in vivo*, and the rapid changes seen after injection of thyroxine, can be shown to be reflections of changes in the sensitivity of mitochondria toward DNP. That sensitivity, as measured *in vitro* by the acceleration of the mitochondrial respiratory rate when increasing amounts of DNP are added, is about half-normal in liver mitochondria from untreated hypothyroid rats, and returns to normal when the rats are pretreated with L-T$_4$ for 3 hours or less, depending on the dose (see Fig. 8–3). The mitochondrial content of hormone rises in parallel with the changes in DNP-sensitivity and respiratory control, but the reversibility of the restoration of DNP-sensitivity (i.e., whether the increased sensitivity depends on the actual presence of the hormone) has not yet been tested.

In hyperthyroid animals, DNP stimulates mitochondrial respiration more than in euthyroid animals. It requires 5 μg per g of L-T$_4$ and a wait of 48 hours to raise the BMR of a euthyroid rat +36 per cent, but then the standard dose of DNP has a calorigenic action of $+90$ per cent, and is lethal. These *in vivo* relationships are summarized in Figure 8–6. Converse to what is seen in hypothyroid rats, the mitochondria isolated from the livers of hyperthyroid rats are more sensitive than normal to DNP added *in vitro*. The time span required for this change, as well as the dose of L-T$_4$, is orders of magnitude greater than in hypothyroid animals.

A tentative mechanism of direct hormonal action can be formulated to accommodate the observations on the thyroid regulation of respiratory control and DNP-sensitivity, and on the stoichiometry of thyroid hormones in the mitochondrion.

In Chapter 5 it was shown that low concentrations of uncoupling agents stimulate respiration *in vitro* or *in vivo* but do not change the rate of phosphorylation. That phenomenon, loose-coupling, has been accounted for by postulating that at low concentrations of an uncoupling agent, the phosphorylation processes have higher affinities for high-energy intermediates, or depend less upon a high proton-motive force, than do the processes depressing electron-transport. The thyroid hormones might have a special role—different from the uncoupling agents—to account for their control of the degree to which uncoupling agents or ADP stimulate respiration. In the chemical theory, Slater (1966b) accounts for respiratory control by postulating that oxidation proceeds via the formation of a nonphosphorylated high-energy intermediate, in which one component, C, "is present in small concentrations compared with the amount of O_2 uptake," i.e., compared with the net electron-flux through the carriers A and B. The paucity of C accounts for the compulsory coupling between oxidation and phosphorylation, because the concentration of C is rate-controlling in respiration (Eq. 8–1).

Equation 8–1 $$AH_2 + B + C \rightleftharpoons A \sim C + BH_2$$
Equation 8–2 $$A \sim C + P_i + ADP \rightleftharpoons ATP + A + C$$
Equation 8–3 $$A \sim C \rightarrow A + C$$

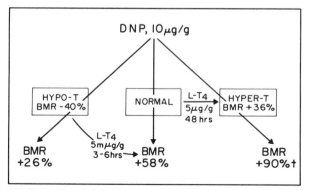

Figure 8–6. The effects of thyroid hormone on the calorigenic action of DNP; a summary of *in vivo* studies. (From Hoch, 1967a.)

When ADP + P_i are absent, respiration is slow (State 4), proceeding only via the liberation of C as in Equation 8–3, by utilization or hydrolysis. When ADP + P_i are present (Eq. 8–2), respiration is rapid (State 3) because C is made available in an energy-conserving reaction to support Equation 8–1. When DNP is present, respiration is rapid (State 3u) because C is made available via stimulation of the energy-wasting route—the process in Equation 8–3.

The synergism between DNP and thyroxine, the dependence of the degree of respiratory control upon the thyroid state, and the paucity of thyroid hormones in mitochondria are accommodated by this chemical theory if one postulates that [C] *depends reversibly upon* [*thyroxine*]. That dependence may range in nature from thyroxine being a component of C, through thyroxine combining with a pro-C to activate it by a conformational allosteric change, to thyroxine reversibly controlling other processes that activate C.

In mitochondria from hypothyroid rats, there is too little thyroxine and too little C; State 4 respiration (Eq. 8–1) will be below normal, and DNP (Eq. 8–3) will stimulate State 4 respiration less than normally. In mitochondria from hyperthyroid rats, there is too much thyroxine, and therefore too much C; State 4 respiration will be above normal, and DNP-stimulation excessive The amount of iodine in normal mitochondria is about one molecule per 50 molecules of cytochrome *a* (i.e., per 50 respiratory assemblies), and the amount of T_4 is one molecule per 200 molecules, fitting the postulate of C as present in controlling amounts. Mitochondria from hypothyroid animals contain only 20 per cent of the normal amount of hormone (i.e., C), whereas thyroxine injection rapidly raises mitochondrial C content 550-fold after a moderate dose. The efficiency of phosphorylation is normal, theoretically and actually, in the presence of low concentrations of hormone. The suggestion that thyroxine acts upon or is actually a component of a chemical intermediate in oxidative phosphorylation has been made before (Martius and Hess, 1951; Lehninger *et al.*, 1959).

There are some difficulties in accepting this hypothesis of an action of thyroid hormones on a chemical intermediate. It is an experimental fact that high concentrations of L-T_4 *in vitro*, and large doses *in vivo*, uncouple mitochondria and lower the P:O ratio. Postulating that [C] depends on [L-T_4] accounts only for loose-coupling and the loss of respiratory control, but not for uncoupling. One must then make a further postulate, that high concentrations of L-T_4 accelerate the hydrolysis of A $\sim$ C, as in reaction 8–3, by a mechanism different from the one that activates C. Perhaps a binding of L-T_4 to additional sites on C might account for a reversal of the activating effects of low concentrations of L-T_4, similar to that seen in some enzyme reactions in which excess amounts of substrates inhibit. Still, it is not comfortable to multiply hypotheses to explain phenomena. With further measurements, it may be preferable to apply our reasoning to the chemi-osmotic hypothesis for a mechanism of action of thyroid hormones; however, the tentative mechanism advanced above serves to rationalize the observations on thyroid hormone actions directly on mitochondria, and perhaps to provide a basis for further experiments. An intriguing area for thought and experiment is the relationship between the direct actions on mitochondrial energy metabolism and the stimulation of protein synthesis by a soluble product of mitochondrial oxidative processes.

Stimulation of protein synthesis via a direct thyroid hormone-mitochondrion interaction has been demonstrated by Sokoloff and his group (Sokoloff *et al.*, 1964, 1968; Sokoloff 1967, 1968) and confirmed by Brown (1966). Their findings are summarized in Figure 8–7.

Adding thyroid hormone to a homogenate prepared from the liver of a normal rat stimulates the ribosomal synthesis of proteins via an acceleration of the processes of translation (whereby peptide chains are elongated by a transfer of amino acids from tRNA-amino acyl complexes). The same phenomenon is seen 2 hours after injecting thyroid hormone. Mitochondria are the locus of action of the hormone both *in vivo* and *in vitro*. *In vitro*, incubation of the mitochondria with a substrate involving the reduction of NAD^+, or with ATP, is necessary (this suggests that an energy-linked process is involved); however, adding ATP, GTP, or reduced glutathione does not replace the effect of hormone-treated mitochondria. When mitochondria preincubated with thyroid hormone are removed from the suspending medium by centrifugation, and the supernatant fluid is heated at 100°C for 5 minutes, and the denatured soluble proteins are spun down, the clear aqueous extract accelerates ribosomal translation to the same degree as intact mitochondria (Fig. 8–8). An acceleration of about +70 per cent is seen with 65 μM L-T_3 and ribosomes from the liver of a starved normal rat. Mitochondria from fed rats produce more stimulation (+200 per cent with 65 μM L-T_4 in our hands). Ribosomes from normal rats respond with a greater acceleration of translation than do ribosomes from hypothyroid rats.

What it is that mitochondria produce to stimulate ribosomal translation—let us call it "Sokoloff's Factor"—is not yet clear. Sokoloff's Factor is water-soluble and does not appear to be either a protein or nucleic acid; it is relatively stable to heat, but is destroyed by ashing (which makes most unlikely its being an inorganic ion, like a Mg salt). Our preliminary studies

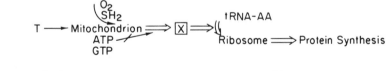

Figure 8–7. The sequence of events after injecting hypothyroid rats with thyroid hormone, (T = L-T_3) or after adding L-T_3 to mitochondria, according to Sokoloff and coworkers. (From Hoch, 1968c.)

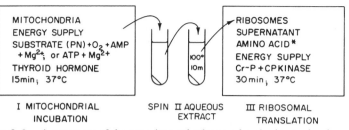

Figure 8–8. A summary of the experimental scheme whereby it can be demonstrated that the incubation of mitochondria with thyroid hormone (I) produces a soluble product (II, called "Sokoloff's Factor" in the text) that accelerates ribosomal translation (III).

(February, 1970) suggest its production depends upon an energy-linked process in mitochondria.

Sokoloff's conclusions and data have been criticized by Tata (1967, 1968), because in Sokoloff's experiments the rates of protein synthesis are said to be only 10 to 15 per cent of those in Tata's system, because the degrees of stimulation are relatively small, and because Tata needed no mitochondria to demonstrate a hormonal stimulation of protein synthesis, but only ribosomes from treated animals. Recent studies of Sokoloff *et al.* (1968) show that such differences arise in part from the protein synthesizing systems chosen, and in part from the experimental conditions. Tata has worked with the depleted systems found in hypothyroid rats; Sokoloff shows that thyroid hormones first and rapidly stimulate protein synthesis in hypothyroid tissues via the mitochondrion and the existing apparatus for protein synthesis (by increasing specific activity), then stimulate the synthesis of more protein-synthesizing apparatus (and increase total activity). When amino acid incorporation is expressed per amount of ribosomal RNA or protein (the baselines Tata *et al.* frequently used), the rates of protein synthesis in Sokoloff's experiments are actually higher than Tata's.

The most recent objections (Tata, 1970) include questions about (1) the specificity of the changes (since thyroid hormones cause mitochondrial swelling, mitochondrial substances may be released that may accelerate ribosomal translation by changing the pH, or ionic strength, or by supplying nucleotides, ions, or other nonspecified rate-controlling factors); (2) the lack of correlation between *in vitro* and *in vivo* findings (although Sokoloff's mitochondrial factor is produced in 5 minutes *in vitro*, the latent period for the stimulation of protein synthesis in Tata's studies is several days); and (3) the contribution of hormone-stimulated mitochondrial protein synthesis. The last two objections are easily countered: Sokoloff has shown very rapid increases in ribosomal translation after *in vivo* administration of the hormone, as well as the slower ones Tata showed, and the observation that an aqueous extract from hormone-treated mitochondria accelerates ribosomal translation removes any chance that mitochondrial synthesis of proteins is being confounded with ribosomal synthesis of proteins. The question as to specificity is answered in part by pointing out that extracts from the mitochondria of hormone-injected animals stimulate translation-like extracts from mitochondria exposed to the hormone *in vitro*; there is no evidence that the mitochondrial extracts have actually altered the pH or ionic strength of the buffered ribosomal systems, or that the hormone accentuates such changes; and it is difficult to see what is nonspecific about the mitochondrion supplying a rate-controlling cofactor to ribosomes, if that is what happens.

A rapid direct action of L-T_3 on liver mitochondria, probably on the synthesis of new mitochondrial DNA-polymerase, in the metamorphosing tadpole of *Rana catesbiana* has been shown by Keir's group (Campbell *et al.*, 1969). The involvement of mitochondria in the primary responses to the hormone appears to be general and not confined to the mammals.

Effects of Thyroid Hormones

An important extramitochondrial effect of thyroid hormones is to control the rate of protein synthesis and perhaps the kind of protein that is synthesized.

The thyroid hormones are necessary for a normal rate of protein synthesis. Liver homogenates obtained from hypothyroid rats synthesize protein slowly (DuToit, 1951; Stein and Gross, 1962). The defect is mainly in the syntheses performed by cytoplasmic ribosomes, although synthesis in the mitochondrion is also depressed.

In hypothyroid rats, one dose of thyroid hormone accelerates cytoplasmic and mitochondrial protein synthesis markedly, but with a long latent period (Tata *et al.*, 1963; Tata, 1964, 1966, 1967, 1968). A relatively small dose of L-T_3, 50 to 175 mμg per g, produces a peak stimulation of protein synthesis by cytoplasmic ribosomes in about 48 hours. The synthesis, as measured by the incorporation of radioactive amino acids into materials precipitable by agents commonly used to precipitate proteins, produces the proteins usually synthesized by the tissue studied, e.g., serum albumin by liver, hemoglobin by bone marrow. In addition, there appears to be a specific relationship between the thyroid hormones and the enzymes of the respiratory chain in mitochondria. Deficiency of thyroid hormones leads to liver mitochondria with an abnormally low content per g of total mitochondrial protein of cytochromes, perhaps flavoproteins, and substrate-specific dehydrogenases. The cytochrome *c* content of tissues begins to fall about 7 days after thyroidectomy (only after 20 days in the myocardium) and reaches a minimum level after 10 days; the basal metabolic rate decreases earlier than the cytochrome *c* content (Klitgaard, 1966). The enzyme deficit is apparently due to a decreased rate of synthesis and is corrected after a single hormone injection, but with a long latent period, up to 90 hours (Freeman *et al.*, 1963; Roodyn *et al.*, 1965). Some of the cytochromes, like the *c* component, are synthesized on the cytoplasmic ribosomes and eventually are bound at the proper functional site in the mitochondrial enzyme complexes; cytochrome c synthesis thus seems to be a special case of the general hormonal stimulation of protein synthesis in the cytoplasm. Some of the other cytochromes, however, and the mitochondrial "structural" protein are synthesized in the mitochondrion in the yeasts and Protozoa, and if mammalian cells operate the same way, a hormonal control may exist over mitochondrial synthesis. Thyroid hormones do stimulate mitochondrial uptake of amino acids into protein when added *in vitro* (Buchanan and Tapley, 1966).

In normal rats treated with thyroid hormones, the rate of protein synthesis changes according to the dose, but even more slowly than in hypothyroid animals. Moderate doses, about 250 mμg of L-T_3 per g, raise the rate of synthesis above normal. Large doses depress the rate below normal. The respiratory enzymes in the mitochondrion increase in concentration; liver mitochondria from hyperthyroid rats contain about twice the normal amount of cytochromes (Drabkin, 1950; Tissieres, 1946).

A well-studied example of the stimulatory effect of thyroid hormones on the synthesis of a specific protein is the *mitochondrial α-glycerophosphate dehydrogenase*. Investigators in Lardy's laboratory (Lardy *et al.*, 1960; Lee *et al.*, 1959; Lee and Lardy, 1965) have shown that the liver mitochondria of rats treated with thyroid hormone oxidize α-glycerophosphate 22 times more rapidly than those of control rats. Activity rises significantly 12 hours after thyroid administration is started, reaches a maximum after 10 days, remains elevated as long as thyroid treatment continues, and returns to normal 10 days after treatment stops. Activity increases in the submitochondrial particles obtained by sonic disruption, indicating that the changes do not depend on increased membrane permeability. Synthesis of new enzyme protein occurs rather than activation of existing enzyme, for inhibitors of protein synthesis prevent the increase in α-glycerophosphate dehydrogenase activity. The thyroid hormone may accelerate synthesis of this enzyme in rat liver mitochondria by directly or indirectly maintaining a stable supply of functioning MRNA (Lee and Miller, 1969).

The newly synthesized enzyme in mice has similar kinetic properties as the enzyme in control tissues, except for a higher V_{max}. In the liver mitochondria of hypothyroid animals, the amount of dehydrogenase is decreased below normal. The increased enzyme synthesis is specific for mitochondria of certain tissues; similar enzymes in the cytoplasm of liver cells and in the mitochondria of brain show no increase in activity. The organ-specific response of α-glycerophosphate dehydrogenase is quite striking, and may resemble the calorigenic responses of different species to the thyroid hormones. Liver and kidney mitochondria of a rat and a mouse, Ehrlich ascites carcinoma cells, and microbial cells show increased activity after thyroid treatment, but the tadpole, frog, chick, and duck show no increase in dehydrogenase activity and no calorigenic effects (Lee and Hsu, 1969; Lardy and Lee, 1961).

The mitochondrial α-glycerophosphate dehydrogenase can function as a component of a system that in effect shuttles reducing equivalents (H-atoms and electrons) from extramitochondrial NADH to intramitochondrial FADH. Mitochondria do not readily oxidize extramitochondrial NADH, and so such systems provide a supply of NAD^+ that allows extramitochondrial oxidations to continue, and they transfer to the mitochondrial phosphorylating system the energy produced by external oxidations. Any metabolite reduced in the cytoplasm to a product which is a substrate for an intramitochondrial oxidation can act as an H-carrier. The soluble cytoplasmic α-glycerophosphate dehydrogenase uses cytoplasmic NADH to reduce dihydroxyacetone phosphate (produced via glycolysis) to L-α-glycerophosphate, which crosses the mitochondrial membrane and is in turn oxidized by the mitochondrial enzyme. The hormonally induced dehydrogenase accentuates the mitochondrial contribution to this cycle, and since mitochondrial oxidations transform much more energy of oxidation into useful forms, the net effect is to emphasize oxidative phosphorylation as well as anaerobic glycolysis. The importance of the normal α-glycerophosphate cycle in vertebrate skeletal muscles has been found to be slight, as it accounts for only about 2 per cent of the rate required to oxidize the cytoplasmic NADH produced by glycolysis (Crabtree and Newsholme, 1969); how much the α-glycerophosphate cycle contributes here after thyroid treatment is not yet known. In isolated perfused rat hearts, the cycle normally appears to be more significant, although lactate oxidation is still limited; in hyperthyroidism, mitochondrial α-glycerophosphate dehydrogenase activity increases three-fold (it is 80 per cent of normal in hypothyroidism) and lactate oxidation is no longer limited (Isaacs, Sacktor, and Murphy, 1969).

There is another hormone-induced synthesis of an enzyme in which changes in nucleic acid metabolism at the levels of transcription and translation have been beautifully demonstrated to occur during the days-long period between administration of the hormone and the first increase of enzyme activity: the carbamyl phosphate synthetase in intact tadpoles and in slices of their livers, investigated in P. P. Cohen's laboratory (Metzenberg et al., 1961; Paik and Cohen, 1960; Kim and Cohen, 1968; Paik et al. 1961; Tatibana and Cohen, 1964; and Shambaugh et al., 1969). This enzyme is of considerable physiologic importance in the establishment of new metabolic patterns in the process of anuran metamorphosis, when the frog begins to synthesize urea rather than excreting ammonia as the tadpole does. However, the extramammalian nature of this system precludes its further discussion here.

Studies with agents that inhibit protein synthesis have given some indication of the mechanisms whereby the thyroid hormones stimulate protein synthesis and produce their calorigenic effect. The administration of puromycin or actinomycin D (which act on ribosomal translation from mRNA and nuclear transcription from DNA, respectively) prevents the

hormone-induced stimulation of protein synthesis, and hormonal calorigenesis *in vivo* (Tata, 1963; Weiss and Sokoloff, 1963). The inhibition of
protein synthesis demonstrates that the hormone affects events in the
nucleus which may be causal in increasing ribosomal synthesis of proteins,
and there is direct evidence for such nuclear changes. The inhibition of
calorigenesis seems to indicate that protein synthesis, either as a general
process or specifically as a synthesis of mitochondrial respiratory enzymes, is
necessary for the rise in BMR; a recent note that puromycin and actinomycin
D prevent DNP-induced, as well as thyroid-induced, calorigenesis throws this
conclusion into some doubt, but requires confirmation (Rossini and Salkho,
1966). (The anticalorigenic effects of inhibitors of protein synthesis are discussed further on page 162.)

The thyroid hormones, when administered to hypothyroid rats, set off
a train of metabolic changes in nucleic acid metabolism that precede and
seem to account for the acceleration of cytoplasmic ribosomal protein
synthesis in the liver. The studies of Tata and his group (Fig. 8–8) have
demonstrated increases first in the metabolism of an RNA-fraction in the
nucleus in 3 to 16 hours, then in the activity of a nuclear DNA-dependent
RNA-polymerase, and in the amounts of soluble RNA (some of which
appears to be ribosomal RNA), then in ribosomal aggregation (about 40
hours). Thus the thyroid hormones, like other hormones (e.g., 17β-estradiol)
that stimulate protein synthesis, first effect increases in nuclear production
of soluble RNA from DNA templates, and then effect increases in the RNAs
that act as messengers or as components of the ribosomes in assembling
peptide chains.

Do the thyroid hormones act directly on the nucleus, or indirectly via
some other part of the cell, to start the train of events leading to protein
synthesis? The evidence at present indicates that the nuclear changes are
not due to a direct action of the hormone. Adding L-T_3 to isolated nuclei
does not stimulate RNA-metabolism or RNA-polymerase activity (Widnell
and Tata, 1963; Tata and Widnell, 1966; Sokoloff *et al.*, 1964). That finding
itself may not rule out a direct action. For example, estradiol added to
nuclei *in vitro* does not produce the same changes as estradiol injected *in vivo*,
probably because a cytoplasmic carrier protein is necessary to carry estradiol into the nucleus. However, estradiol appears in the nucleus after *in
vivo* injection, and the thyroid hormone does not (Dillon and Hoch, 1967).

However, studies on cultured human kidney epithelial cells (Siegel and
Tobias, 1966) have provided evidence that added thyroid hormone localizes
at the nuclear genome or membrane. By radioautography, up to 80 per cent
of the hormone is found after as little as one hour of incubation, in the peri-
nuclear zone and on the nucleus but not in the nucleoli. The functional
concomitants of this apparent localization are not clear; although actinomycin D and puromycin both inhibit the formation of colonies by these
growing cells, thyroid hormones reverse the effects of puromycin (which are
exerted upon ribosomal translation) but not those of actinomycin D (on

nuclear transcription). On the other hand, actinomycin D blocks the localization of the hormone in the nuclei. The resolution of these radio-autographs ($\times 900$) was not high enough to show the location of the mitochondria.

The fact that the thyroid hormones control the amount of *all* the cytochromes in mitochondria can be taken as evidence that the thyroid hormones act on the mitochondrion and not on the nucleus, if we assume that the genomes in mammalian cells are distributed as in yeasts and Neurospora (the circularity of mammalian mitochondrial DNA does indicate a similarity to that of unicellular organisms). In yeasts and Neurospora, there is convincing evidence that the mitochondrial genome codes for cytochromes *a* and *b* and for the structural protein, and that the mitochondrial protein-synthesizing apparatus translates for the actual synthesis. The nuclear genomes code for mitochondrial cytochrome *c*, and the synthesis is performed on the endoplasmic reticulum. If thyroxine acts on the mitochondrion, it should control only cytochromes *a* and *b*, but not *c*. The control over cytochrome *c* synthesis might be exerted through the mitochondrion via Sokoloff's Factor or via a hormonal control over the synthesis of the mitochondrial structural protein that permits only a certain number of cytochromes *a*, *b*, and *c* to be "seated" properly. Only when the mitochondrion is considered the site of hormone action can the equal changes in *all* the cytochromes be rationalized.

The evidence against a primary action of the thyroid hormones on the nuclear genes that controls protein synthesis may be summarized as: (1) thyroid hormones do not change nuclear RNA metabolism when added to isolated nuclei; (2) thyroid hormone injected *in vivo* does not arrive in nuclei; and (3) a nuclear action does not account for the hormonal control over mitochondrial content of cytochromes *a* and *b*, but only cytochrome *c*

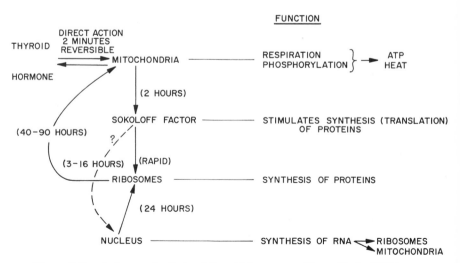

Figure 8–9. Actions and effects of thyroid hormone. (From Hoch, 1970.)

(assuming mammalian cells are like Neurospora and yeast in their genetic control over mitochondrial cytochromes). The early nuclear changes in Tata's studies thus seem to be secondary to a primary hormone action elsewhere. Sokoloff's demonstration of a primary action in the mitochondrion remains to be related to the observed nuclear changes.

It now appears possible to integrate the actions and the effects of the thyroid hormones as in Figure 8–9.

Small doses of the hormone act rapidly or instantaneously, and reversibly, on mitochondria. The functional changes in mitochondria accelerate, by a process not yet clear, the rate of translation of tRNA-amino-acyl complexes by ribosomes to synthesize proteins. Among the proteins synthesized are the enzymatic components of the mitochondrial respiratory chain. Increased protein synthesis is an effect of the early mitochondrial changes and a cause of the late mitochondrial changes. The nucleus is also involved early, but the relation between the rises in nuclear RNA-metabolism and the earlier changes in mitochondrial composition is not yet finally defined. Thus, mitochondria show changes in function dependent upon the hormone's presence or absence, and change in enzyme content secondary to the alterations in protein synthesis that depend ultimately upon the functional changes.

In Vitro Actions of Thyroid Hormones

It has been an axiom for workers in the field of hormone action that before a mechanism is acceptable it is necessary to show that the hormone added in "reasonable" concentrations *in vitro* acts just like the hormone injected *in vivo*. That axiom has lost some force recently. For example, there does not seem to be much doubt that the hormone 17β-estradiol stimulates protein synthesis by acting primarily and directly on the cell nuclei in target tissues such as the uterus, based on demonstrations that injected estradiol rapidly reaches those nuclei. Yet 17β-estradiol added to isolated uterine nuclei *in vitro* has no such action, probably because a specific cytoplasmic protein is necessary to introduce the hormone into the nucleus in an active form (Gorski *et al.*, 1968).

The thyroid hormones have direct actions on mitochondria *in vitro* that resemble those seen after hormone injection *in vivo*. L-T_4 or L-T_3 depresses the respiratory control of isolated mitochondria and stimulates State 4 respiration provided that the proper experimental conditions are met. Those conditions include an assay mixture that contains cytochrome *c* and NAD^+, little or no Mg^{2+} ions, and no bovine serum albumin. The first two ingredients are necessary for demonstrating increased State 4 respiration, and the last two must be minimized because they are antagonists of thyroid hormone action.

The concentrations of thyroid hormones that are effective in accelerating respiration *in vitro* are usually between 5 μM and 50 μM. Interestingly, the same range of concentrations stimulates protein synthesis *in vitro* in Sokoloff's systems. Some investigators have objected to ascribing any physiologic significance to these *in vitro* experiments, because the concentrations of hormone are many orders of magnitude above what might be expected to exist *in vivo*, i.e., about 0.1 μM to 0.001 μM. Such objections seem valid, but there may be other reasons for requiring so much hormone in the *in vitro* system besides its "unphysiologic," pharmacologic effect. For one reason, the studies on 17β-estradiol, alluded to above, show that a carrier protein is required to introduce that hormone into the nucleus in active form. Similar "activation" might be required for thyroid hormones, so that only a small fraction of the added dissolved thyroid hormone reaches a mitochondrial site. For another reason, a concentration of 50 μM inside the mitochondrion is easily attained after injection of L-T$_4$ *in vivo* (see Fig. 8–4), and so is not really "unphysiologic." Lastly, there are mitochondrial components that do change in function in the presence of as little as 0.1 μM L-T$_4$ *in vitro*.

The energy-dependent reductions of intramitochondrial pyridine nucleotides are very sensitive to thyroid hormones *in vitro*. L-T$_4$, 0.1 to 0.5 μM, inhibits the rate and the extent of reduction of pyridine nucleotide upon addition of succinate (Chance and Hollunger, 1963c; Hess and Brand, 1964), a process that is the result of the reversal of electron-transport in Complex I. It is necessary to limit the concentration of added Mg^{2+} salts to demonstrate these hormonal actions. The significance of such findings, in addition to the relatively high sensitivity toward the hormone, is that the "bottle-neck" in respiration is thought to be at or near the pyridine nucleotide portion of the electron-transport chain, probably involving the interactions between the substrates and the dehydrogenase enzymes. The hormones thus act *in vitro* upon a rate-controlling site. Because the hormonal changes were effected in the presence of oligomycin, it was concluded that they involved decreases in the amounts of nonphosphorylated high-energy intermediates. Such an action would be energy-wasting (corresponding to the process in Eq. 8–3, p. 94) and like that of DNP, but not like that of injected L-T$_4$ *in vivo*. It remains to be seen whether energy-conserving actions of thyroid hormones involve the pyridine nucleotides or other components.

The thyroid hormones do not act on mitochondria *in vitro* exactly like DNP, although there are many similarities. Thyroxine makes mitochondria swell and DNP does not; DNP blocks the swelling action of thyroxine (see Lehninger, 1962). Thyroxine and DNP act synergistically upon mitochondrial oxidative phosphorylation when added together in concentrations that separately have no effects. All these observations indicate that thyroxine and DNP have different mechanisms of action. The biologic specificity of the hormone is most evident *in vivo*: thyroxine, but not DNP, relieves the defects in hypothyroidism, stimulates growth and development, and induces metamorphosis in the *Anura*.

In connection with the demonstrations of the *in vitro* sensitivity of mitochondrial pyridine nucleotides to thyroxine, it may be significant that the thyroid state *in vivo* controls the amount of pyridine nucleotides in rat liver mitochondria. In normal rats, injection of thyroid hormones halves the total pyridine nucleotide content of liver mitochondria. In hypothyroid rats, the liver mitochondria contain twice the normal amount of pyridine nucleotides. These changes are in the opposite direction to those in the mitochondrial content of cytochromes and to the changes in the BMR under thyroid control. Klingenberg (1963) has suggested that the rate of electron-transport in the pyridine nucleotide region of the chain depends on the balance between the opposing tendencies of pyridine nucleotide binding to the substrate-dehydrogenase enzymes (which slows the rate) and to the pyridine nucleotide oxidases. In our terms, hypothyroidism favors binding to the dehydrogenases rather than to electron-transport components, and hyperthyroidism favors the opposite.

A special role for the thyroid hormones in the metabolic processes involving intramitochondrial pyridine nucleotides is suggested by *in vitro* studies on the *mitochondrial pyridine nucleotide transhydrogenase*. Mammalian mitochondria from many tissues contain an enzyme that catalyzes the transfer of a proton plus two electrons between NAD^+ and $NADP^+$:

Equation 8–4 $NADH + NADP^+ \rightleftarrows NAD^+ + NADPH$

This enzyme, the pyridine nucleotide transhydrogenase, is apparently associated structurally with the inner mitochondrial membrane, and can be found in the oxidizing and phosphorylating subparticles prepared by fragmentation of mitochondria (Ball and Cooper, 1957; Kaplan *et al.*, 1953, 1956; Devlin, 1959). A broad biochemical significance for the function of the PN-transhydrogenase is implied by the fact that it is also found associated with the respiratory system of some bacteria (Colowick *et al.*, 1952).

Mitochondria obtained from several tissues, but especially liver, contain firmly bound NAD^+ and $NADP^+$. In general, the oxidation of NADH in mitochondria proceeds along intramitochondrial electron-transport pathways and results in the generation of utilizable energy-rich bonds or ion gradients; the oxidation of NADPH proceeds via extramitochondrial pathways that are involved in reductive biosyntheses. The process in Equation 8–4, catalyzed by the PN-transhydrogenase, thus represents a switching point where reducing equivalents obtained by the oxidation of substrates are routed toward energy-conserving or synthetic ends. In mitochondria, there is evidence that NADH is very rapidly formed or oxidized upon addition of substrate or ADP, but NADPH is much more slowly formed or oxidized (Klingenberg & Slenczka, 1959). It appears, then, that $NADP^+$ receives or donates H^+ plus two electrons only via the PN-transhydrogenase reaction, and that the specific activity of the transhydrogenase is significantly less in normal mitochondria than the net activity of the reactions that transport electrons along respiratory pathways.

The reaction shown in Equation 8–4 has been called the "classic" transhydrogenase reaction; it does not involve the transfer of high-energy bonds, and the NAD^+-NADH couple has a redox potential only about 50 mv more negative than the $NADP^+$-NADPH couple. However, ATP (or the energy derived from ATP) can increase and uncoupling agents can decrease the amount of NADPH formed in Equation 8–4. ATP does not appear to be an obligatory component of the transhydrogenase reaction, as it is not used up stoichiometrically, but its influence is referred to by the terms "energy-controlled," "energy-linked," or "energy-dependent" transhydrogenase reactions. That influence is similar to the ATP-induced reversals of electron-transport and the intramitochondrial accumulation of ions. Among the pieces of evidence that energy-controlled transhydrogenation, reversal of electron-flow, and oxidative phosphorylation involve the same high-energy nonphosphorylated intermediate are observations that the first two processes are inhibited by oligomycin when they are driven by ATP but are not inhibited by oligomycin when they are driven by the energy derived from substrate (always pyridine-nucleotide dependent) oxidation; that they are inhibited by thyroid hormones to the same degree; and that the PN-transhydrogenase competes with oxidative phosphorylation for the energy of the nonphosphorylated intermediate. The data on energy-linked PN-transhydrogenation are discussed by Klingenberg, Ernster, Estabrook, Hommes, and others in a 1963 symposium (see Chance, 1963).

Relatively low concentrations of thyroxine (0.1 to 1 μM) inhibit the partly purified "classic" transhydrogenase of beef myocardium *in vitro* (Ball and Cooper, 1957), and somewhat higher concentrations inhibit transhydrogenase activity in intact mitochondria and in the particles obtained from mitochondria after treatment with digitonin (Devlin, 1959; Hommes and Estabrook, 1963). The effective concentrations of thyroxine approach those presumed to exist in the tissues of euthyroid animals. The thyroid hormones may thus exert a direct action on this enzyme that controls a branch point in energy metabolism. To interpret the physiologic significance of these *in vitro* experiments, it would be well to know the effects of hyperthyroidism and hypothyroidism. Although in a preliminary report the transhydrogenase activity in the tissues of thyroid-treated animals was found to be normal (Stein *et al.*, 1959), recent studies on mitochondria show changes in the redox kinetics of the pyridine nucleotides that are consistent with inhibition of the transhydrogenase after injection of L-T_4 (Hoch, unpublished). It is not yet clear whether the changes observed represent an alteration in the specific activity or in the amount of transhydrogenase, or involve energy-controlled or energy-independent processes.

Another way to deduce the influence of thyroxine upon this enzyme is to measure the concentrations of the oxidized and reduced forms of these pyridine nucleotide coenzymes, and to assume that the intramitochondrial

PN-transhydrogenase is the controlling factor. Kadenbach (1966) has shown that the energy-dependent reduction of NAD$^+$ and NADP$^+$ upon the addition of succinate is similar in the mitochondria obtained from the livers of either normal rats or rats injected 4 days previously with thyroid hormone. No data on earlier effects of the hormone are on hand.

The significance of the PN-transhydrogenase reaction for understanding the actions of the thyroid hormone is not yet clear. Kaplan *et al.* (1956) stressed the fact that the pathways whereby NADPH is oxidized are non-phosphorylative, and are energy-wasting, heat-producing processes that compete both with syntheses that require NADPH and with the energy-conserving oxidative phosphorylation of NADH:

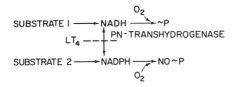

Figure 8-10. The action of the pyridine nucleotide transhydrogenase in connecting phosphorylative processes with nonphosphorylative processes. Thyroxine inhibits transhydrogenase activity.

When thyroxine inhibits transhydrogenase activity, the reducing equivalents obtained by the oxidation of substrates (2) via NADP$^+$ cannot be shunted to the phosphorylating NAD$^+$ pathway, but proceed along the nonphosphorylating NADP$^+$ pathway. This represents a partial uncoupling. It is also possible that the actions or effects of the thyroid hormones on the PN-transhydrogenase system are one manifestation of a more general involvement of the hormone in the formation of the (hypothetical) high-energy non-phosphorylated intermediates.

D-THYROXINE

The biologic specificity of the changes that follow the administration of the naturally occurring hormones has been examined through the use of compounds chemically related to L-thyroxine but lacking physiologic activity. D-thyroxine has been so used, because it is thought to have no calorigenic or other physiologic effects. Actually, that is true only when the dose of D-T$_4$ is low and comparable to that of L-T$_4$. Higher doses are calorigenic; D-T$_4$ has been said to lower serum cholesterol concentrations without raising the BMR, and thus to be of use therapeutically in persons with hypercholesterolemia, especially when they have evidence of impaired coronary circulation (Starr, 1964). Because the doses necessary are close to those that are calorigenic, and because calorigenesis raises the cardiac demands for oxygen, the use of D-T_4 is not without risk.

D-T_4 is about as effective as L-T_4 on mitochondrial oxidative phosphorylation *in vitro*, unlike its lesser effect on calorigenesis *in vivo*. D-T_4 administered *in vivo* does reach the liver mitochondria, with almost as much iodine appearing there as after a similar dose of L-T_4 (Hoch, unpublished results). However, D-T_4 injected *in vivo* does not stimulate protein synthesis in the mitochondrion-dependent systems Sokoloff studies, whereas D-T_4 added *in vitro* is almost as effective as L-T_4 (Sokoloff, 1967). He suggests that D-T_4 might be more rapidly removed from the mitochondria *in vivo* than L-T_4; however, our preliminary data show elevated iodine contents as long as 3 hours after D-T_4 injection. Since such mitochondria do not stimulate ribosomal translation, it appears that the biologic specificity of the hormone isomers might reside in their effects on the production of Sokoloff's Factor. Such a conclusion may cast some doubt on the specificity of the direct actions of L-T_4 on mitochondrial oxidative phosphorylation in relation to control over protein synthesis. The data at present seem consistent with the tentative hypothesis that D-T_4 exerts the same direct actions as L-T_4 but not the delayed effects of L-T_4; testing D-T_4 for rapid actions on mitochondrial function seems appropriate for rejecting or accepting the hypothesis. Part of the difficulty is that D-T_4 is not an inert, inactive form of the hormone, and its effects must be judged from quantitative rather than qualitative differences from those of L-T_4. When analogues of L-T_4 or L-T_3 are used, rather than the stereoisomer, interpretation is difficult because the part of the hormone that is substituted may have been involved in its mechanism of action.

General References: Several reviews (Hoch, 1962a, b; 1968) cover the ideas on thyroid hormone mechanisms presented here. This is still a very controversial area. Other views are discussed by Tata *et al.* (1962), Tata (1964), Wolff and Wolff (1964), Sokoloff (1968), Tapley (1962, 1964), Hillmann (1961), and Kadenbach (1966).

9

CONTROL OF ENERGY TRANSFORMATIONS: THYROID HORMONES

In considering how the thyroid hormones control energy metabolism, and how excess or deficiency of hormone changes energy metabolism to produce the symptoms and signs of disease, it is useful to draw an analogy between the mitochondrion and a machine. That analogy, as first used by Lipmann and Green and other early workers in the field, likens the mitochondrion to an energy-transducer that burns fuel and so liberates energy, and converts the liberated energy to perform mechanical, electrical, chemical, or osmotic work.

The performance of such a mitochondrial machine can be estimated from (1) its *power*, in units of the rate of production of useful energy ($\sim P/t$); (2) its *rate of fuel consumption*, the input of total energy per unit of time (O/t); and (3) its *efficiency*, the ratio of output of useful energy per input of total energy ($\sim P/O$). The power term measures the machine's capacity to do work, and represents the product of the efficiency and the fuel consumption.

A diagram of the performance of the mitochondrial machine in terms of its power, fuel consumption, and efficiency, as affected by variations in the amount of thyroid hormones, is shown in Figure 9–1. In hypothyroidism, fuel consumption is abnormally slow, but efficiency is normal; power, the

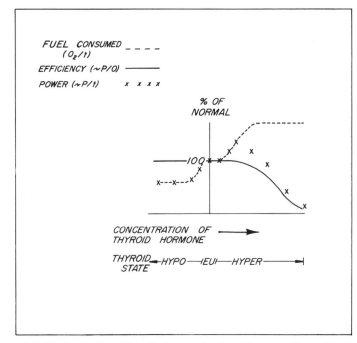

Figure 9–1. Oxidative power in relation to the thyroid state. (From Hoch, 1970.)

capacity to do work, is diminished. Increasing the amount of thyroid hormone (as in treating hypothyroid subjects) accelerates fuel consumption still at normal efficiency, until power is normal. With further increases of thyroid hormone (mild hyperthyroidism) there is a further acceleration of fuel consumption above normal levels but still with no decrease in efficiency, resulting in power greater than normal. At high levels of thyroid hormones (severe hyperthyroidism), no further respiratory acceleration may occur, but efficiency decreases; depending on the degree of inefficiency, power is normal or subnormal. Such a scheme would predict *biphasic effects* of administered thyroid hormones, depending upon the dose; small doses would be energy-conserving and increase the capacity for supporting energy-requiring processes, and large doses would be energy-wasting. The effect of administered hormone would depend upon the thyroid state of the organism. A dose of hormone that, given to a hypothyroid subject, raises oxidative metabolic power to normal levels, might in a euthyroid subject decrease power.

The resting hyperthyroid subject thus consumes oxygen faster because the mitochondria in his tissues consume oxygen faster in State 4. The term "hypermetabolism" in this connection is a misnomer, for while oxidation is increased, the metabolic processes of phosphorylative energy transfer are normal or decreased. This kind of hyperoxidation arises from the presence of excess amounts of thyroid hormones in the mitochondria. Not all organs consume more oxygen in thyrotoxic subjects; the brain, spleen, and testes do not. The mitochondria of brain, spleen, and testes do not swell in the presence of added thyroid hormones, in contrast to those from other tissues (Tapley and Cooper, 1956), but it is not known if their iodine contents are high or normal in thyrotoxic subjects. Another kind of hyperoxidation arises from the late adaptive increases in mitochondrial number and enzyme content, but it seems to be fully efficient and should persist for some time even when the amount of thyroid hormone in the mitochondria becomes normal.

Faster oxidation produces heat faster. The oxidative rate is probably the chief determinant of heat production, even when the efficiency of the mitochondrion is normal. Moderate increases in heat production can be compensated by sweating, vascular dilation, and flushing, with resultant tachycardia, increased cardiac stroke volume, and pulse pressure. Such changes may in part be mediated through the action of the hormones of the adrenal medulla (see Chap. 11). Compensation may be precarious, and the excessive demands imposed by relatively slight increases of external temperature, the rise of heat production during muscular exercise, or administration of agents with uncoupling properties may exceed the physiologic adjustments to dissipate heat and produce fever. In the most extreme forms of thyrotoxicosis, in which uncoupling may occur and evolve more heat, the fewer may be catastrophic (as in a thyroid crisis).

Conversely the hypothyroid subject consumes oxygen more slowly than normally, because his mitochondria respire slowly in State 4. Both because of the deficiency of thyroid hormone and the depletion of respiratory enzymes, "hypometabolism" is here not a misnomer; phosphorylative metabolism proceeds at a low rate because of the diminished liberation of oxidative energy. Decreased rates of oxidation produce less heat, and extremes result in hypothermia (see myxedema coma, below). Demands for increased heat production, as in acclimatization to cold environments, are met poorly by hypothyroid subjects. Responses to the hormones of the adrenal medulla are subnormal. Clinical investigators (Selenkow and Marcus, 1960) have commented upon the "apathetic" hyperthyroidism seen in older patients, that has features similar to those of hypothyroidism; on our basis, both are reflections of the decreased capacities for the production of utilizable energy.

METABOLIC CHANGES

The defects in the transformation of energy in clinical and experimental hypothyroidism and hyperthyroidism may underly some of the metabolic changes observed. In general these diseases may be thought of as manifestations of deranged energy-transfer. The hypothyroid subject is a penurious fuel-saver, whose slow oxidation limits the rate of production of useful energy and of heat. The hyperthyroid subject is a fuel-waster, producing excess heat and, in extreme cases, little useful energy. "For some reason the appetite and absorptive powers in hyperthyroidism are not quite sufficient to maintain weight, although laboring men with a much higher caloric requirement per day remain in nutritive equilibrium" (DuBois, 1916). For a more detailed discussion, the reader is referred to reviews of this area (Hoch, 1962b, 1968c).

Synthesis of Macromolecules

The synthesis of macromolecular compounds, which uses amino acids, hexoses, or acetate to form proteins, glycogen, or lipids respectively, depends upon an energy supply, and is controlled by thyroid hormones. The rates of synthesis of macromolecules are depressed in hypothyroidism. The catabolism of the macromolecules is also depressed, usually more than the synthesis, so that a net increase in storage occurs. The basis of the slow catabolism in hypothyroidism is not quite clear; it may be due to depressed synthesis of the hydrolytic enzymes or to depressed demand. The predominance of depressed catabolism is most striking in the metabolism of cholesterol; the hypercholesterolemia of hypothyroidism is due to depressed excretion even though the synthesis is retarded.

Administration of thyroid hormones to normal or hypothyroid subjects produces biphasic effects on the rate of synthesis of proteins, lipids, and glycogen. Relatively low doses stimulate the synthetic rate, and high doses depress it; these biphasic effects are consistent with the model for production of useful energy discussed on page 109. It should be noted, in judging whether a dose of hormone is large or small, that the more hypothyroid a subject is, the higher his sensitivity to thyroid hormones; this has already been mentioned in connection with calorigenesis. Doses of L-T_4 of the order of 10 to 100 mμg per g stimulate the synthesis of fats, proteins, or carbohydrates in extremely hypothyroid subjects but have no measurable effects in normal subjects. In hypothyroidism, doses higher than the stimulatory range depress protein or cholesterol synthesis, for instance.

In thyrotoxicosis, or in normal subjects treated with thyroid hormones, the rate of synthesis of proteins, lipids, and glycogen again show biphasic effects depending on the severity of the disease or the dose of the hormone. Glycogen synthesis appears to be especially sensitive, and is usually depressed in human thyrotoxicosis, although in experimental animals the hormone can stimulate synthesis. Thyroxine also accelerates oxidations and weight loss, and wasting of both fat and lean body mass (Wayne, 1960) occurs in thyrotoxicosis without any losses in appetite. The hypocholesterolemia that is commonly seen usually reflects a stimulated excretion that supervenes over increased synthesis.

Coenzyme Syntheses

The synthesis of the water-soluble vitamins and at least one fat-soluble vitamin (vitamin A) into coenzymes or other biologically indispensable molecules is controlled by the thyroid hormones. The synthetic steps under hormonal control are usually energy-requiring condensations and phosphorylations. Thyrotoxic subjects often have conditioned vitamin deficiencies, in which normal intake of vitamins is accompanied by deficiency symptoms because of increased demands or defective utilization or both. Among the water-soluble vitamins, thiamine, pyridoxine, cobalamin, ascorbic acid, and pantothenic acid deficiencies, and defects in the tissue metabolism of their corresponding coenzymes, have been described.

Vitamin A metabolism is affected biphasically. Both hypo- and hyperthyroid patients have poor dark adaptation. In hypothyroidism, serum vitamin A is decreased because carotene is not converted to the vitamin; hormone treatment restores synthesis. In euthyroid animals, the hormone increases vitamin A synthesis, but prolonged treatment produces a severe, resistant vitamin A deficiency.

Muscle Contraction

The dependence of muscle contraction and relaxation upon ATP is reasonably well-established, although the chemical events are still not defined. Probably because of the fact that muscle mitochondria rapidly resynthesize muscle energy-rich $\sim$P-bonds (in ATP and in phosphocreatine, see the next section) after their use in contraction and relaxation, the defects observed in thyroid disease seem directly attributable to changes in energy transformations. In human hypothyroidism, skeletal muscles are larger and firmer than normal and contract slowly because of an abnormality in the contraction mechanism. The clinical sign of the "hung-up" reflex, with its slow relaxation, and the characteristic severe muscle cramps reflect this defect. "Hypothyroid myopathy" is accompanied by a lowered activity of muscle α-glucosidase, which is restored after treatment with thyroxine (Hurwitz *et al.*, 1970); no reports of mitochondrial function are on hand. In hyperthyroidism, muscle contracts at the normal rate but performs work inefficiently. In addition, there is atrophy of muscle fibers, perhaps owing to defective protein synthesis. Clinically "thyrotoxic myopathy" reflects these abnormalities. Administering thyroxine has a biphasic effect on muscular work, a low dose improving the work done per contraction and a four times higher dose decreasing it (Ganju and Lockett, 1958).

There is, however, some controversy now about whether skeletal muscle mitochondria from thyrotoxic human subjects show any abnormalities at all; Peter (1968) and Dow (1967) have found normal respiration in both State 4 and State 3, and normal respiratory control in mitochondria prepared from biopsy specimens.

Those reports must be considered in the light of earlier studies by Johnson *et al.* (1958) which showed chemical evidence (low ATP and high ADP + P_i concentrations) for uncoupling of mitochondrial oxidative phosphorylation in the skeletal muscles of thyroxine-treated rats; other studies on human patients apparently similar to those of Dow and of Peter show loose-coupling, i.e., high State 4 respiration (Ernster *et al.*, 1959), or abnormally high rates of respiration in both State 4 and State 3 (Stocker *et al.*, 1966, 1968). Part of the variety in the results may arise from the reversibility of the direct actions of thyroid hormones, because bovine serum albumin was used in the preparation and assay in some of the studies. However, the more recent work of Peter (1968) has taken this into account, and bovine serum albumin was omitted. The resolution of this confusing story is not yet apparent, but there are certain general observations that may be made. The data on skeletal muscle mitochondria might be explained several ways.

(1) Skeletal muscle does not contribute to the calorigenic effect of the thyroid hormones. The hormone content of isolated skeletal muscle mitochondria is not as responsive to the thyroid state as is that of liver mitochondria in rats. Skeletal muscles contain about half the iodine found in the normal body, but muscle mitochondria contain less iodine per g of protein than liver mitochondria (Table 9–1). Muscle mitochondria obtained from hypothyroid rats contain 70 per cent as much iodine as those from normal rats, and injected hormone raises that value to normal, but neither change is statistically significant. Liver mitochondria contain 20 per cent of the normal amount of iodine, and the same injection of L-T$_4$ raises that content to 60 per cent; both are significant differences.

Although skeletal muscle represents about 40 per cent of the total body mass, it is conceivable that the thyroid hormone raises the BMR by acting mainly on the visceral organs, especially the liver. The liver is said to contribute up to 30 per cent to the BMR; to raise the BMR by 50 per cent, liver respiration would have to rise about 2.7-fold, an increase well within the capacity of mitochondrial State 4 respiration. Furthermore, there is

Table 9–1 *Iodine in Mitochondria from Normal and Hypothyroid Rats ($\mu g\ I/g\ protein$)* *

	LIVER MITOCHONDRIA	P	MUSCLE MITOCHONDRIA	P
I Normal rats	0.43		0.31	
II Hypothyroid rats	0.10	<0.001 vs I	0.22	<0.2 vs I
III Hypothyroid rats + L-T$_4$, 5.2 mμg/g, 2 m	0.24	<0.025 vs II	0.33	<0.2 vs II

* From Dillon and Hoch, 1967.

reason to believe that the liver contributes even more than 30 per cent to the BMR (see Chap. 6), whereas skeletal muscle consumes much less oxygen than its 40 per cent contribution to body mass would lead one to predict (see p. 66).

(2) Skeletal muscle does contribute to the calorigenic effect of thyroid hormones, but they way skeletal muscle mitochondria are prepared and assayed *in vitro* obliterates the abnormalities that exist *in vivo*. The low hormone contents in the isolated mitochondria and the questionable rise in iodine after hormone treatment (seen in Table 9–1) may be due to the preparative method. The findings of Johnson *et al.* (1958), alluded to above, were made on intact total skeletal muscle without any fractionations and are consistent with mitochondrial uncoupling. Even when bovine serum albumin is omitted from the preparation and assay, there is still reason to believe that the behavior of skeletal muscle mitochondria *in vitro* does not reflect their condition *in vivo*. The procedure of isolating skeletal muscle mitochondria may cause selection of loosely coupled mitochondria when the homogenization is incomplete, although tightly coupled mitochondria can be obtained by more vigorous homogenization of the same tissue (Hulsmann *et al.*, 1967). The homogenization and suspension of these mitochondria is always performed in solutions containing Mg^{2+}, ATP, KCl, and sucrose (in contrast to liver mitochondria, which are prepared readily in sucrose solutions). Mg^{2+} ions are known antagonists of thyroid action. Further, the skeletal muscle contains considerable amounts of nonmitochondrial protein that may extract mitochondrial thyroid hormones (Dillon and Hoch, 1967). Lastly, even when a known uncoupling agent, DNP, is injected *in vivo* in a dose that raises the BMR several-fold and is lethal, the mitochondria that are isolated from skeletal muscle respire completely normally *in vitro* (Hoch, unpublished data), although mitochondria from liver are loose-coupled (Hoch, 1968b). It appears that the last word on skeletal muscle mitochondria and thyroid hormones has yet to be said.

Heart muscle mitochondria, on the other hand, are particularly susceptible to thyroid hormones (Bing, 1961). Clinically this seems to be reflected in the high incidence of myocardial failure in thyrotoxicosis (the increased work load and decreased efficiency of contraction are an unfortunate combination), and in the relatively poor response of this form of failure to digitalis (which is more effective against mechanically induced defective contractile mechanisms).

Creatine Metabolism

In muscle, the $\sim$P-groups of phosphocreatine represent a storage form for energy-rich bonds; phosphocreatine, as shown in Figure 9–2, can generate ATP from ADP through the action of a transfer enzyme, creatine phosphokinase, in an iso-energetic reaction The amount of $\sim$P stored in this form in muscle is enough to support only a few contractions through the

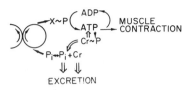

Figure 9–2. Creatine (Cr) and inorganic phosphate ion (P$_i$) metabolism in relation to mitochondrial energy transfer. Creatine phosphate (Cr ~P), in equilibrium with the generated ATP, in severe hyperthyroidism. The extra free Cr and P$_i$ are excreted. The P$_i$-pool is augmented through inefficiency of oxidative re-esterification. Similarly, administering extra Cr in a load test augments the Cr-pool and subsequent excretion. (From Hoch, 1968c.)

formation of ATP, and must itself be replenished from mitochondrial energy transformation via ATP.

In all muscle diseases, alterations in creatine metabolism are common. In hyperthyroidism, creatine excretion increases and in hypothyroidism it decreases. Injection of normal subjects with thyroxine rapidly depletes muscle phosphocreatine content, and more slowly increases creatine excretion (Wang, 1946), again a finding consistent with a hormonal action on muscle mitochondrial oxidative phosphorylation. The defect in muscle creatine metabolism in thyrotoxicosis was early described as an inability to "fix" creatine, and was demonstrated either with administered or endogenous creatine (Shorr *et al.*, 1933–34; Thorn and Eder, 1946). The abnormally high excretion of creatine in the "load-test" is so characteristic that it has been used diagnostically. In more modern terms, the inability to "fix" creatine arises from a defect in the resynthesis of phosphocreatine from ATP and creatine, perhaps through a deficiency in ATP generation. Such a rationale for the observations is shown in Figure 9–2.

Inorganic Metabolism

Phosphate metabolism is altered by the thyroid state. Hyperthyroid subjects are in negative phosphate balance (Rawson *et al.*, 1955) and lose P$_i$ in the urine; the phosphaturia does not involve the parathyroid glands, but is a response to thyroid hormones. The soft tissues of thyrotoxic animals esterify inorganic phosphate slowly, and contain more P$_i$ and less ATP than normally. Both the soft tissues and bone turn over phosphate abnormally rapidly, in bone probably in conjunction with changes in calcium ion metabolism. Hypothyroid subjects treated with the hormone promptly excrete large amounts of phosphate because of increased dissociation or hydrolysis of phosphocreatine, but this may be a transient phenomenon caused by demand.

Calcium ions are excreted excessively in the urine and feces in hyperthyroid patients, causing a negative calcium balance; hypercalcemia is occasionally seen; again, parahormone is not involved. Hyperthyroidism is accompanied by an accelerated Ca^{2+} turnover in bones, which is slowed to normal when treatment restores euthyroidism. In hypothyroid rats, Ca^{2+} incorporation into bone is retarded.

Magnesium ion balances are positive in hyperthyroidism and negative in hypothyroidism. The metabolic changes that account for these abnormalities are not yet fully understood, and there is still some disagreement about experimental results. In hyperthyroidism, plasma magnesium content is low (not all workers agree that it is), but in spite of the positive balance the total and cellular exchangeable Mg^{2+} is reported to be normal (Jones *et al.*, 1966). Administration of the hormone to myxedematous patients results in the rapid urinary excretion of large amounts of Mg^{2+}, although the cellular Mg^{2+} is very low in untreated hypothyroidism. The effect of administering magnesium salts upon thyrotoxicosis is controversial, some finding a decrease in BMR and heart rate (Hueber, 1939), others finding no decrease in BMR or change in negative nitrogen and phosphate balances (Wiswell, 1961); one wonders if Mg^{2+} ions reverse the actions but not the effects of thyroid hormones. The role of Mg^{2+} in energy transformations is further discussed in Chapter 14.

Decreased exchangeable *potassium* and hyperkalemia and hyperkaluria occur in thyrotoxicosis. Periodic muscular paralysis, with postprandial hypokalemia, is associated with hyperthyroidism, especially in the studies in Japan where the first syndrome is not uncommon.

Thyroid Storm and Myxedema Coma

The thyroid hormones are involved in the control of body temperature mainly through their control of the rate of oxidation. Hyperthyroidism raises heat production by increasing the rate of oxidation, and in severe cases by decreasing the efficiency of energy conversion. Usually the excess heat can be dispelled by physiologic compensations, such as flushing, sweating, and increased circulation; many of the clinical characteristics of hyperthyroid patients arise from these compensations, which are mediated in large part through the accentuated actions of the catecholamines. *Thyrotoxic crisis* or storm can be viewed as a failure of compensation due to increased heat production through a loss of mitochondrial efficiency. Body temperatures rise sharply to 107° or more, muscle tone is lost, liver damage (perhaps, mitochondrial) is severe, and the patients may die (Bansi, 1939; Pemberton, 1936; Lamberg, 1959; Waldstein *et al.*, 1960; Harrison, 1968; McArthur *et al.*, 1947). Body refrigeration may remove enough heat to save the situation, but therapeutic measures to provide adrenocortical hormones and to antagonize adrenomedullary hormones have also been used with success.

The clinical picture of thyroid storm is very similar to that seen when thyroid hormones and uncoupling agents are administered together (see Chap. 13). Administering large doses of thyroid hormones to animals usually produces a loss of weight and an apathetic death, not a hyperthermic crisis. Much smaller doses of hormone, however, can produce a fatal hyperthermia

in conjunction with the action of an uncoupling agent that acts on mitochondrial oxidative metabolism.

Hypothyroidism depresses heat production by diminishing the rate of
mitochondrial oxidation. Physiologic compensations preserve body heat.
The skin is cold, circulation is slow, and cold is poorly tolerated; the actions
of the catecholamines are suppressed. Body temperatures may be below
normal. Infections that normally elicit fever may not raise the hypothyroid
patient's temperature at all, or at least not above normal. Occasionally a
fatal hypothermia may supervene, the so-called *myxedema coma*, in which body
temperature can no longer be maintained, and has been reported as low as
74°F (Nielsen and Ranløv, 1964; Nickerson *et al.*, 1960; Forester, 1963;
Leon-Sotomayor and Bowers, 1964). Experimentally the calorigenic
response of hypothyroid rats to an administered uncoupling agent is subnormal, because their mitochondria are subnormally sensitive.

Effects of Other Hormones and Drugs

A number of hormones and drugs act more or less intensely, depending
upon the thyroid state of the subject, as has been mentioned above in connection with specific systems. In general, hyperthyroidism exaggerates and
hypothyroidism minimizes the changes seen after administration of the
agent. Physiologically the clearest example is that of the *catecholamines*. The
relationship is so striking that some have concluded that the apparent peripheral effects of thyroxine are actually effects of epinephrine (Brewster
et al., 1956), but there is evidence against so sweeping a claim (see Hoch,
1962b).

The dependence of the calorigenic effect of epinephrine upon the thyroid state is an example of the generality that the thyroid state controls the
response of the body to calorigenic substances. Excessive rise in BMR is
seen in hyperthyroidism, and little or no rise is seen in hypothyroidism, after
the administration of glucagon (Davidson *et al.*, 1960), nitrophenols, salicylates (Hoch, 1965a), chlorpromazine (Hoch unpublished data; Skobba
and Miya, 1969), and "febrile toxins" (above). The only exception seems to
be the enhanced sensitivity of hypothyroid subjects to the thyroid hormone
itself. For a theoretical basis for the dependence of calorigenesis upon thyroid state, see page 94.

Other Features

There are several clinical features of hyperthyroidism and hypothyroidism that are not readily reduced to manifestations of changed cellular
energy-transfer. These features fall into two categories. In the first category
are those features that arise from mechanisms not owing directly to the
changed amounts of thyroid hormones in the tissues but to phenomena
associated with the primary defect in thyroid hormone production. Thus

exophthalmos is one of the classical signs in the Merseburg triad in hyperthyroidism, but it is not produced by hormone administration (Means *et al.*, 1963), and frequently persists after euthyroidism is restored.

In the second category are those clinical features that may arise from changed amounts of tissue thyroid hormone, but that are not reducible to cellular phenomena because of insufficient information at present. The hyperirritability of the nervous system in hyperthyroidism, and its opposite in hypothyroidism, are often presenting symptoms clinically. The involvement of ATP in nerve conduction and in resynthesis of acetylcholine at the myoneural junction and the involvement of K^+, Ca^{2+}, and Mg^{2+} in neural events make it likely that hormone-induced defects in energy-transfer and ion-accumulation will affect the nervous system, but just how changes in nerve function relate to "nervousness" presents complex problems not yet conclusively approached. Similarly, the presenting abnormality of "myxedema" seems to be a defect in mucopolysaccharide metabolism that leads to excessive deposition, somehow dependent upon insufficient thyroid hormone. The thyroid hormones obviously control growth, development, and the striking structural and chemical changes in anuran metamorphosis. In general terms we may say these processes must depend upon available "energy," but our lack of knowledge of the details of the energy-dependent steps precludes a mechanistic interpretation at present.

10

CONTROL OF ENERGY TRANSFORMATIONS: CATECHOLAMINES

The structure of the catecholamines (Fig. 10–1) is a simple one; they are substituted di-orthophenols.

The purpose of the catecholamine actions are, in an old and time-tested phrase, "to prepare the animal for fight or flight." In terms of energy transformations they mobilize sources of rapidly utilizable energy from the body's storage depots by promoting the breakdown of glycogen and lipids to glucose, and fatty acids and glycerol. The appearance of large amounts of oxidizable substrates is accompanied by a burst of increased oxidation— i.e., by the rapid utilization of the mobilized energy sources.

The rapidity and short duration of their actions make the catecholamines "hormones of opportunity" that set off immediate changes to meet presenting conditions. The stimuli that see off the discharge of catecholamines are usually in the category of immediate demands: sudden stresses, fear, and rage. A more sustained secretion of catecholamines also occurs

Figure 10–1. The structure of the catecholamines.

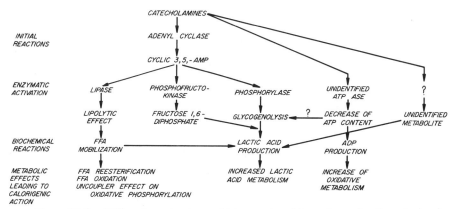

Figure 10–2. Different mechanisms which may be of importance for the calorigenic action of the catecholamines. (Modified from Lundholm *et al.*, 1966; by permission of the authors and The Williams and Wilkins Company, Baltimore.)

in response to prolonged exposure to cold. The short-term energy-mobilizing responses of the catecholamines have been contrasted with the longer-range energy-mobilizing responses of the thyroid hormones that have been thought to represent the setting of a governor or thermostat to control the "idling speed" of the organism. The two sets of energy-controlling hormones are complementary and interdependent. The degree of response to the catecholamines depends on the actions of the thyroid hormones.

The catecholamines have calorigenic actions on animals ranging from the frog to man. Indeed, in his review aptly titled "Fact and Theory Regarding the Calorigenic Action of Adrenaline" (1951), Griffith attributed the very term "calorigenic action" to Boothby and Sandiford in 1922, following as they said a suggestion of Lusk as a description of the increase in the respiratory metabolism of animals and man that is caused by the administration of epinephrine.

A great deal has been written about the calorigenic action of the catecholamines, but it must be admitted that the mechanism whereby they act is not yet completely understood. There is not even any agreement that they raise the rate of oxygen consumption by an action that involves only one set of subcellular processes. Some investigators, including Griffith (1951), consider this calorigenic response "not as a single unitary reaction but as the integrated sum of several." In general, the proposed actions fall into several categories, which are outlined in Figure 10–2.

ACTIONS ON MITOCHONDRIAL RESPIRATORY CONTROL

Catecholamine calorigenesis has many of the features of nitrophenol calorigenesis, and has been thought to be exerted via a direct action on mitochondria similar to the direct action of DNP. After administration of a catecholamine, O_2 consumption rises rapidly, reaching a peak in 20 minutes, and gradually decreasing for the next hour (Fig. 10–3).

There is evidence for a direct uncoupling action of the catecholamines. Sobel *et al.* (1966) injected mice subcutaneously with L-epinephrine, L-norepinephrine, or L-isoproterenol (5 μg of base per g), killed them 6 hours later, and examined the mitochondria prepared from their hearts. The P:O ratios were lower than normal, but the respiration, respiratory control, swelling, and ATPase activity were normal. The P:O ratio was depressed *pari passu* as the myocardial catecholamine content increased, and this was thought to provide a basis for impairment of cardiac function. Pretreatment with propranolol, a β-adrenergic blocker, enhanced the increase in myocardial catecholamines and the depression of the P:O ratio; pretreatment with dibenzyline or reserpine inhibited both the rise in myocardial epinephrine and the drop in P:O ratio.

The sort of uncoupling seen here, without a rise in State 4 respiration, is unusual. It does not account for the early calorigenic action, but that rise would be missed by waiting for 5 hours after catecholamine injection. No early rise in mitochondrial respiration at a time when the calorigenic action is at its peak has been demonstrated in Sobel's or others' studies. Nor could Sobel *et al.* (1966) and a number of earlier workers show any effects of catecholamines added *in vitro* to rat heart or other mitochondria. Indeed, the P:O ratio of liver mitochondria 30 to 45 minutes after injection of epinephrine is reported to be normal, and the mitochondria appear more resistant to deterioration upon standing (Lianides and Beyer, 1960b). The evidence for a direct calorigenic action of catecholamines upon mitochondrial respiratory control is thus quite uncertain. Nor does an uncoupling action

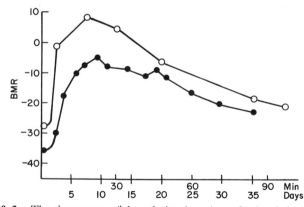

Figure 10–3. The time course of the calorigenic actions of epinephrine and thyroxine. The per cent change in the BMR of two human subjects with myxedema is plotted against time after the intravenous injection of 10 mg of DL-thyroxine (●—●) and subcutaneous administration of 0.5 mg of epinephrine (○—○). The time scale for thyroxine is in days and for epinephrine in minutes. The figure does not demonstrate the latent period of action of thyroxine. Values are replotted from the original data of Boothby and Sandiford (1922). (From Tata, 1964; by permission of the author and Academic Press, New York.)

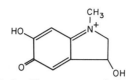

Figure 10–4.   The structure of adrenochrome.

of the catecholamines, which would decrease the capacity of the organism to do work because of metabolic inefficiency, make much sense if we accept a constructive purposeful role for the catecholamines in meeting environmental demands.   Still, the idea is hard to dismiss, and it is supported by a number of pieces of indirect evidence.

Adrenochrome is an oxidation product of epinephrine (Fig. 10–4). Adrenochrome is calorigenic when administered *in vivo;* it also produces psychotic behavior.   Unlike epinephrine, adrenochrome acts as an uncoupling agent when added to mitochondria *in vitro* (Park *et al.*, 1956; Krall *et al.*, 1964).   The calorigenic action of epinephrine thus might arise after its oxidation to adrenochrome.   Low concentrations of adrenochrome and thyroxine, which are ineffective when added to mitochondria separately, produce uncoupling when combined, and such observations suggest a basis for the observed dependence of epinephrine-induced calorigenesis upon the thyroid state of animals.   However, the failure to detect much, if any, adrenochrome in tissues seriously weakens these hypotheses.

In hyperthyroid subjects, the catecholamines raise the metabolic rate more than in normal subjects, and in hypothyroid subjects, less or not at all. The thyroid state controls in a similar manner the calorigenic actions of other endogenous or exogenous agents.   The common basis for these phenomena may be a thyroid hormone control of mitochondrial responses (see Chap. 8).   The other actions of the catecholamines, the glycogenolytic and hyperglycemic, the lipolytic, and the intropic, all depend as well upon the thyroid state.   A common basis for these interdependences has been thought to lie in thyroid hormone effects on the metabolism of the catecholamines, or on biochemical steps which are common to the mechanisms whereby the catecholamines exert their varied actions (Harrison, 1964).

The thyroid hormones appear to control the amounts or the specific activities of a number of enzymes that affect catecholamine actions.

(1) The thyroid hormones inactivate the enzymes that normally inactivate the catecholamines; in this manner, excess thyroid hormone would lead to excess tissue catecholamine concentrations.   The catechol-O-methyl-transferase, the amine oxidases, and a peroxidase system have been shown to be under such thyroid influence, but it is still difficult to assign physiologic relevance to the specific mechanisms.   It is also unclear whether the tissue contents of catecholamines vary in a way that is consistent with this thyroid action.

(2) Thyroid hormones control the amount of adenyl cyclase, the enzyme in the adrenergic cascade (Fig. 10–5) that transforms ATP to cyclic 3,5-AMP (CyAMP) in adipose tissue (Krishna *et al.*, 1968). CyAMP is the "second messenger" for the catecholamines that mediates most of their effects, and an increased production of CyAMP (as in hyperthyroidism) would account for increased catecholamine effects. It would be convenient if CyAMP itself were a respiratory stimulant. Unfortunately, adding CyAMP to rat liver mitochondria does not accelerate respiration (Hoch, unpublished observations), nor does injecting anesthetized rats with up to 100 μg per g intravenously or intraperitoneally raise the BMR (Strubelt, 1968), nor does injecting theophylline (which inhibits the phosphodiesterase that rapidly inactivates CyAMP) before CyAMP demonstrate any BMR increase. However, on the reasoning that injected CyAMP does not enter cells well, Strubelt has adduced the demonstrated calorigenic action of theophylline itself to support the idea of a calorigenic role for CyAMP. Small doses of methylxanthines are thought to act only via the catecholamines, but larger doses may have a direct calorigenic action (Strubelt and Siegers, 1969). Caffeine depresses the ∼P content of muscle (Miyazaki, 1962) in consistence with an accumulation of CyAMP at the expense of ATP.

It is also of interest that CyAMP has been thought to be involved in the mechanism of muscle contraction (Mommaerts *et al.*, 1963); that might account for the inotropic effects of the catecholamines on the myocardium. As will be seen in the next section, the production of free fatty acids and lactate via CyAMP-induced lipolysis and glycogenolysis may stimulate mitochondrial oxidation by supplying substrates. Some investigators believe that the fatty acids, which are uncoupling agents *in vitro*, are also uncoupling agents *in vivo* and thus accelerate oxidations by a direct action.

(3) Thyroid hormones control the specific activity of the mitochondrial enzymes that respond to uncoupling agents and ADP and that produce ATP

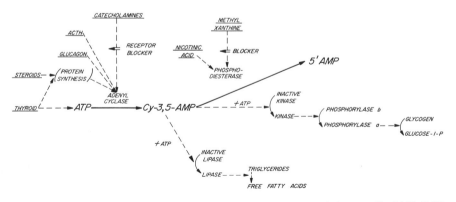

Figure 10–5. The "adrenergic cascade." ATP is converted into cyclic 3′,5′-AMP (Cy 3,5-AMP) through the catalytic action of adenyl cyclase. The scheme shows the effects of Cy 3,5-AMP on lipolysis and glycogenolysis, the inactivation of Cy 3,5-AMP by conversion to 5′-AMP, and the effects of several hormones and drugs.

(the sole source of CyAMP). The effects of thyroid hormones on catechol-amine calorigenesis and DNP calorigenesis are very similar: hypothyroid animals respond to both agents subnormally, administration of thyroid hormone increases the calorigenic response to each within 3 to 6 hours (Bray and Goodman, 1965; Hoch, 1965a), and hyperthyroid animals have exaggerated calorigenic responses. The calorigenic responses to DNP *in vivo* are reflected in similar hormone-sensitive mitochondrial responses *in vitro*. By analogy, the alterations in calorigenic responses to catecholamines might arise from hormone-induced mitochondrial changes as well. (It is too bad that argument by analogy is not very convincing.) Thyroid hormones affect catecholamine calorigenesis very much like they affect DNP calori-genesis, and this may be taken as further indirect evidence for a direct action of catecholamines on mitochondrial respiratory control, since the calori-genic action of DNP is exerted on mitochondria. The thyroid hormone sensitizes animals to the actions of the catecholamines within a few hours (Bray and Goodman, 1965), indicating that the presence of thyroid hormone is necessary rather than the synthesis of new mitochondrial respiratory as-semblies (which takes days). Further, the catecholamines increase the BMR more in thyroid-treated subjects than in normal ones, and that is accounted for only by an increase in mitochondrial sensitivity, but not in the number of mitochondria (Hoch, 1965a).

The "uncoupling" theory of catecholamine action has received some support from studies on preparations of isolated fat cells. In lipocytes ob-tained from the epididymal fat pads of rats, about half the normal respiration is not linked to phosphorylation, as shown by the inhibitory action of oligo-mycin on respiration and phosphorylation; epinephrine accelerates the respiratory rate in the presence and in the absence of oligomycin. Hepp *et al.* (1968) inferred that the calorigenic effect of epinephrine involves respiration not coupled to phosphorylation. In brown fat pads obtained from hamsters, normal respiration is controlled mainly by ADP and so appears to be more closely coupled to phosphorylation than it is in epididymal fat; norepinephrine stimulates respiration in brown fat in a "mixed" manner, i.e., the initial phase being ADP-controlled, and the postmaximal phase being loose-coupled (and no longer ADP-controlled) possibly through direct ac-tions of the free fatty acids that have been liberated (Prusiner *et al.*, 1968c).

The interpretation of findings with fat cells is complicated by the fact that the free fatty acids are also the major substrate for oxidation and require an energy-supported activation first; that makes ambiguous the observed effects of inhibitors and uncoupling agents. Studies on isolated rat hearts (Challoner and Steinberg, 1965) seem to confirm the findings obtained with adipose tissues. In such hearts, contraction can be blocked with 31 mM K^+ in the perfusate (eliminating contributions of contraction-produced ADP to respiratory stimulation). Epinephrine stimulates resting respiration markedly, and oligomycin then has little inhibitory effect, again indicating the presence of a high level of nonphosphorylating respiration.

Indeed, these workers showed similar results with beating hearts, but others have attributed the epinephrine effect there to increased production of ADP.

Williamson *et al.* (Fisher and Williamson, 1961; Williamson, 1964, 1966; Williamson and Jamieson, 1966) have described a sequence of events in perfused rat hearts during the first 90 seconds after addition of epinephrine. First, the amount of CyAMP increases and the force of myocardial contraction increases; the connection between the two may be causal, or epinephrine may affect contraction and CyAMP production separately. Contraction is at the expense of ATP, and so the ADP concentration rises. Then ADP stimulates mitochondrial respiration (State 3), and thus accounts for the calorigenic effect. Glycogenolysis and increased supplies of substrates from carbohydrate sources are rather late and do not occur in time to account for calorigenesis. It is not clear at present just how these results are reconcilable with those showing loose-coupling.

There is no apparent proof here that the respiration is phosphorylative, and the ADP increase may be a result of loose-coupling or uncoupling, or a shift of steady state reflecting the increased utilization of ATP, rather than the cause of respiratory stimulation. On the other hand, it is not certain that oligomycin added to a perfusate penetrates to all the respiring cells in an isolated organ, and a lack of penetrance, with an ensuing lack of inhibition, would appear as nonphosphorylating respiration; studies with cell suspensions show much more extensive inhibition by oligomycin.

It is possible, as suggested by the earlier workers and supported by Prusiner's recent work, that the catecholamines stimulate respiration by different mechanism successively. If this is true, it is a phenomenon analogous to that seen with the thyroid hormones; both groups of energy-controlling hormones seem to produce physiologic changes that serve the purpose of utilizing more energy by employing different mechanisms successively.

SUBSTRATE SUPPLY

The mechanisms for catecholamine action discussed in this chapter so far have concerned changes in mitochondrial respiratory control. Many workers feel that the calorigenic actions of the catecholamines arise through an indirect stimulation of cellular respiration via an increased supply of substrates. Certainly the catecholamines, via the "adrenergic cascade," liberate and make available oxidizable substrates.

A disturbing feature of substrate-supply theories is a theoretical one. If the calorigenic action of the catecholamines is indeed caused by increased amounts of substrates being made available for oxidation, we must assume that the oxidizing apparatus in the tissues before the introduction of catecholamines was functioning in a starved state, needing only more fuel to cause it to oxidize at a more nearly maximal rate. But all we know of mitochondrial function in the living cell (see Chap. 6) indicates that mitochondria respire slowly because of intrinsic metabolic control mechanisms, not because of deprivation of substrates. Yet there are observations—that infused carbohydrates or lipids are calorigenic—that seem inconsistent with the data on mitochondria. In rabbits, infusing enough lactate to raise the blood lactate level to the same degree as after epinephrine, raises oxygen consumption as much as epinephrine does (see Lundholm *et al.*, 1966). In man, infusing that much lactate raises oxygen consumption only half as much as epinephrine does (Svedmyr, 1966a and b). In dogs pretreated with adrenal demedullation and exposed to cold, lactate does not raise the metabolic rate. Nicotinic acid, however, which selectively blocks the rise of blood fatty acid level, lowers but still allows a major portion of the calorigenic effect of norephinephrine in man and in rabbit. Low doses of nicotinic acid that block the lipolytic effect of epinephrine, but not its effect on blood lactate, have no influence on the calorigenic effect either. It is difficult to say which is cause and which is effect in these experiments. As Lundholm *et al.* put it, "an increase in the oxidation of FFA can certainly take place as a result of an increased oxygen consumption but it is uncertain whether an increase in plasma FFA causes increased oxygen consumption."

It seems at present that norephinephrine and epinephrine can raise the metabolic rate by mechanisms other than their lipolytic or glycogenolytic effects, but that lipolysis is somehow involved. Mechanisms invoking the production of ADP from ATP by means other than uncoupling have been suggested. Thus a catecholamine stimulation of muscular movements and contractions would use up ATP, produce ADP, and thereby raise mito-chondrial respiration (as discussed above). But the blockage of muscle contraction with curare or other agents does not abolish catecholamine calorigenesis. A burst of re-esterification of fatty acids and glycerol, after the catecholamine-induced lipolysis, would produce ADP from ATP, but there is evidence that this does not occur to any extent that accounts for the amount of oxygen uptake.

Summary: The catecholamines increase the metabolic rate rapidly and transiently. The mechanism of this action is not understood, nor is it clear that there is only one mechanism. There is evidence for direct actions of the catecholamines upon mito-chondrial respiratory control and for indirect effects mediated by increased con-centrations of ADP or free fatty acids. Lipolysis and fat metabolism seem to play a major role, perhaps causally or perhaps to supply energy. The calorigenic action (as well as the glycogenolytic, lipolytic, and inotropic actions) depends upon the thyroid state of the subject, being excessive in hyperthyroidism, and low or absent in hypothyroidism; that dependence may involve mitochondrial responses, or steps in the adrenergic cascade, or both of these.

CHAPTER

11

CONTROL
OF ENERGY
TRANSFORMATIONS:
OTHER
HORMONES

The actions and effects of a number of hormones, as far as they are under-stood at present, appear to involve processes in energy metabolism. This is hardly surprising. It has been thought for some time, because of the effectiveness of hormones in such small amounts, that some sort of amplifying mechanism must mediate their control over the transformations or trans-locations of much larger numbers of molecules (e.g., Green, 1941). Two processes were singled out that could serve as amplifying systems: the specific activity of enzymes and the permeability of membranes. Membrane permeability, as has been discussed in the sections on mitochondria, is now known to involve energy-transducing mechanisms in many cases rather than being simply a phenomenon of molecules or ions passing through holes. In addition, some of the enzyme activities that are hormonally controlled involve steps in energy transduction.

INSULIN

Insulin is a calorigenic hormone. In rabbits injected with 0.5 units of insulin per kg of body weight, the metabolic rate rises +26 per cent above basal levels in the first hour. The calorigenic effect depends upon the thyroid state of the animal, and in rabbits pretreated with thyroxine, insulin raises the metabolic rate to almost $+80$ per cent above basal (Ryer and Murlin, 1951). The increased utilization of oxygen apparently involves not only the increased oxidation of glucose, but of lipids and proteins as well.

There is a considerable amount of evidence that insulin affects mito-chondrial metabolism. When rats are injected with insulin labeled co-valently with ^{131}I, the hormone rapidly localizes in the mitochondria and microsomes of the liver, where it is bound very firmly (Lee and Williams, 1954; Lee and Wiseman, 1959). No studies on the normal endogenous content of insulin in mitochondria are on hand and the physiologic signifi-cance of this localization of exogenous insulin is somewhat dimmed by the questions as to the effects of insulin on the liver. Glucose enters liver cells, in contrast to other tissues, without the need for insulin; however, protein synthesis in the livers of rats is affected by insulin *in vivo*. In diabetic rats, amino acids are incorporated more slowly into proteins by liver mitochondria or microsomes than in normal rats, and insulin restores the synthesis; the liver preparations do not respond to insulin *in vitro*, although skeletal muscle does (Korner, 1960). These findings raise the possibility that insulin-induced calorigenesis is another example of a specific dynamic action.

While there is some question about the localization of insulin in mito-chondria, it has been found in a number of laboratories that insulin affects oxidative phosphorylation and mitochondrial respiration. Early findings of Shorr *et al.* (1940) suggested defects in the turnover of phosphate in the tissues of diabetics. These findings seemed to be explained by observations that mitochondria obtained from the livers of diabetic rats or cats show depressed phosphorylation and respiratory control and are few in number, larger and more fragile than in normal liver (Vester and Stadie, 1957; Hall *et al.*, 1960; Hohorst *et al.*, 1963; Katsumata and Ozawa, 1969; Schaefer and Naegel, 1968a, b). Only severely diabetic animals demon-strate these differences, and while some workers observe uncoupling of oxidative phosphorylation with alloxan-treated rats, others do not and find it necessary to use pancreatectomized cats. Depressed respiration (State 3) of the liver mitochondria occurs in both of these animal preparations. Since no measurements of State 4 respiration are reported, it is not possible to correlate these mitochondrial changes with the metabolic rates. In a very small series of biopsies of skeletal muscles from human diabetic patients, the isolated mitochondria respire slowly in both State 3 and State 4 (Lundquist and Svanborg, 1964); Makinen and Lee (1968), in contrast, report normal phosphorylating efficiency, high respiratory control, normal cytochrome content, and a twice-normal activity of succinate oxidase in mitochondria prepared by a different method.

The administration of insulin to the diabetic animal restores the mito-chondrial efficiency of phosphorylation and the rate of respiration. In rat State 3 and State 4 respiration are rapidly accelerated, and the ATP:ADP ratio increases in 90 minutes. These, being seen so soon after hormone administration may be direct actions of insulin. A difference set of mitochondrial changes (effects of the hormone, perhaps) occurs from 2 to 3 days after single or repeated injections of insulin, which includes a restoration of the depleted number of mitochondria, and of the NAD^+-dependent dehydrogenases involved in the oxidative catabolism of glucose. The net changes are thought to reflect influences of insulin on the relative rates of energy-controlled H-fluxes in the mitochondrial flavoprotein $\rightarrow NAD^+$ system, or in the pyridine nucleotide transhydrogenase system $NADH \rightarrow NADP^+$.

The effects of insulin on mitochondrial oxidative phosphorylation have been demonstrated most usually after injection *in vivo*. In an intact organ (perfused rat hearts), adding insulin to the aqueous perfusate does not alter oxygen uptake, although DNP does (Fisher and Williamson, 1961). These hearts were doing no work, and the penetrability of a large molecule like insulin might be questioned; the authors warned against relating their finding to heart metabolism *in vivo*. Some workers have found that insulin added *in vitro* alters mitochondrial function. Insulin, together with other hormones like oxytocin and vasopressin that contain a disulfide bond, and glutathione, causes mitochondria to swell when added *in vitro* in concentrations around 10^{-5} M (Lehninger and Neubert, 1961). ATP reverses this swelling. It was suggested that the hormone's disulfide groups altered mitochondrial membrane permeability. More recent studies (Cash *et al.*, 1968) indicate that impurities, such as Zn^{2+} ions and possibly other metals, that are found in all but specially purified preparations of insulin are responsible for the action on the mitochondrial membrane. Pure insulin does not make mitochondria swell. Most attempts to demonstrate uncoupling *in vitro* have failed, even when high and obviously unphysiologic concentrations of insulin were added. For instance, the mitochondria of pig heart phosphorylate and respire normally in the presence of 1.5×10^{-5} M pig insulin (Leblanc *et al.*, 1968).

There does not seem to be much doubt that severe diabetes and insulin deficiency are associated with defects in mitochondrial function and composition, and that insulin administration restores normality. Whether the defects and restoration represent direct actions of insulin on the mitochondrion is less clear. Insulin has widespread influences, and the mitochondrial changes might be secondary to one of them; for instance, the absence of insulin promotes lipolysis, and the actions of free fatty acids on mitochondria might account for many of the observations. More studies on mitochondrial insulin content seem indicated.

GLUCAGON

This polypeptide hormone secreted by the pancreas induces rapid glycogenolysis, which raises blood glucose levels. Glucagon acts via cyclic

3,5-AMP much like the catecholamines, but only on the liver (see Foa and Galansino, 1962), except at pharmacologic levels.

Glucagon is a calorigenic hormone (Davidson *et al.*, 1960). In rats, subcutaneous doses between 0.135 and 5 μg per g raise the metabolic rate linearly with the logarithm of the dose. The calorigenic action is rapid and transient, reaching a maximum about 1 hour after injection, and lasting about 1 hour. When the dose is increased above 5 μg per g, no further rise in the BMR occurs; the maximal calorigenic effect is about +50 per cent. Such maximal calorigenesis might arise if the calorigenic agent acted on only one or a few organs. If glucagon's calorigenic action is exerted only on the liver, as its glycogenolytic action is, and if the liver contributes about 30 per cent to the total BMR, a five-fold stimulation of liver respiration would account for the calorigenic effect reaching a maximum of only +50 per cent.

The calorigenic action of glucagon depends upon the thyroid state. In thyroidectomized rats, glucagon does not raise the BMR; 5 to 10 μg of thyroxine per day restores the response. The calorigenic action or effect of glucagon also depends on the temperature of the environment to which rats are acclimated. In rats kept at 22°C, glucagon produces the usual rapid transient rise in BMR, but in rats kept at 10° or 15°, no additional rise in BMR occurs after glucagon injection above the elevation produced by the exposure to low temperatures (Holloway and Stevenson, 1964). In hypothermic rats, glucagon does not produce hyperglycemia (Crawford *et al.*, 1965).

The lack of glucagon-induced calorigenesis suggests that whatever agent raises the BMR in cold acclimation acts at the same target as and not additively with glucagon. If the cold acclimation agent is thyroxine (see Chap. 15), glucagon is the only calorigenic compound that is not hyperactive in the hyperthyroid subject; on the other hand, cold acclimation produces changes that are not completely characteristic of hyperthyroidism.

The mechanism of the calorigenic action of glucagon is not known. The hyperglycemic effect is not necessary for calorigenesis; glucagon is calorigenic in fasted animals but does not raise their blood sugar. There is evidence that glucagon stimulates tissue respiration directly. Incubating liver slices with glucagon *in vitro* increases respiration by +15 per cent to +50 per cent (Davidson *et al.*, 1960; Suzuki, 1961), and injecting rats intravenously with 0.05 μg of glucagon per g 15 minutes before sacrifice stimulates the respiration of liver slices +50 per cent above control levels (Suzuki, 1961). These respiratory effects fall short of the five-fold stimulation of liver respiration required to account for the calorigenic effect. Glucagon also affects phosphorylation, perhaps like an uncoupling agent. Prolonged glucagon treatment depletes liver ATP but not muscle ATP, showing either the consumption of ATP for the resynthesis of liver glycogen (Giacovazzo *et al.*, 1959), or a decrease in ATP synthesis. Glucagon, like parathormone, seems to act directly on the kidney to increase the P_i in the urine (Barac, 1965); however, glucagon is not an uncoupling agent when added to rat kidney mitochondria (Lefebvre and Lelievre, 1964) or to rat liver mitochondria (unpublished observations), and it produces no changes in respiration or phosphorylation. Thus, although the dependence of glucagon calorigenesis on thyroid hormones is somewhat similar to that of insulin or uncoupling agents, it does not appear at present that glucagon acts on mitochondria directly like an uncoupling agent.

PARATHYROID HORMONE

The parathyroid hormone is a single polypeptide chain (molecular weight about 9000) that alters the metabolic functions of bone, kidney, and intestines. With recent developments in radio-immunoassay and methods of tissue culture, studies on the mechanism of action of this hormone have become more precise (Levell and Fourman, 1968). In general, it acts on the translocation of ions. There are apparently two independent and complementary actions: a slow and persistent action on bone that mobilizes Ca^{2+}, and a rapid and transient action on the kidney tubule that results in increased concentrations of P_i in the urine. Parathormone also is required, together with vitamin D, for the absorption of calcium by the small intestine. The changes in bone and kidney metabolism seen after *in vivo* administration of the hormone are also produced by *in vitro* additions of parathormone, but not the intestinal changes. The actions of the hormone on ion translocations in isolated mitochondria seem to account for the renal and bone actions, and are the reason for a discussion of parathyroid hormone in connection with cellular energy fluxes.

Parathormone can act as an uncoupling agent *in vitro* (Dawson and Jones, 1957), and *in vivo* (Whitehead and Weidmann, 1959). *In vitro*, the hormone inhibits oxidative phosphorylation in homogenates of liver and kidney, as measured by the uptake of $^{32}P_i$. Ca^{2+} ions also inhibit, and at levels below the inhibitory range, Ca^{2+} potentiates the action of parathormone. The inhibition by parathormone or Ca^{2+} ions is reversed by increasing the concentration of Mg^{2+} ions (Aurbach *et al.*, 1965; Coh *et al.*, 1966; Sallis and DeLuca, 1966). *In vivo*, intraperitoneal injections of 200 USP units of parathormone per kg per day for 3 days depress the growth rate of kittens, although they continue to feed as well as the control animals; this suggested to Whitehead and Weidmann a general effect on metabolism. Single doses of parathormone markedly (45 per cent) decrease $^{32}P_i$ uptake into ATP in cartilage cells in actively ossifying areas of bone and act like injections of DNP. There is no record of a demonstration that parathormone is calorigenic, as it should be if its effect on liver, kidney, bone, and perhaps other tissues is similar to that of an uncoupling agent.

There are two other actions of the hormone on mitochondria: (1) The release of calcium phosphate from mitochondria previously loaded with these ions; this is an artificial *in vitro* phenomenon, but suggests a general increase in the permeability of membranes for the diffusion of calcium phosphate. The hormone does not act this way on mitochondria from vitamin D-deficient animals, but addition of vitamin D releases calcium in the absence of hormone (DeLuca *et al.*, 1962). (2) The hormone stimulates the uptake of the K^+ or Mg^{2+} salts of phosphate by mitochondria, and the release of H^+ ions, by diverting oxidative energy into pumping phosphate ions. Ca^{2+} ions can also be accumulated. The relatively high concentrations of hormone that are necessary and the fact that mitochondria from liver (which is not a target tissue) are acted on as well as those from kidney are objections to the specificity of these actions *in vitro*.

These two sets of observations on mitochondria have been invoked to explain the hormonal actions on both kidney and bone, and some of the apparently secondary effects. In addition, cyclic AMP has been implicated as an intermediate in the actions of parathormone; the hormone activates renal and skeletal adenyl cyclase *in vitro* (Chase *et al.*, 1969). Synthesis of protein and an energy supply are apparently necessary for parathormone action on bone in tissue culture and on plasma calcium levels in live animals; both are blocked partly or completely by actinomycin D and puromycin (Rasmussen *et al.*, 1964; Tashjian *et al.*, 1964; Kunin and Krane, 1965). O_2 is necessary and DNP inhibits (Goldhaber, 1963, Martin *et al.*, 1965). The effects of actinomycin D and puromycin are difficult to visualize if they act only on protein synthesis and parathormone acts only on ion translocation.

PITUITARY HORMONES

Heat production decreases in hypophysectomized mammals. One obvious cause is that the absence of the thyrotropic hormone produces a secondary hypothyroidism; however, observations that heat production declines more abruptly after hypophysectomy than after thyroidectomy, that the administration of thyroid hormones raises the BMR more in an athyroid than in an apituitary animal, and that thyrotropic hormone does not restore the BMR to normal levels in hypophysectomized animals, all indicate that the hormones of the pituitary have calorigenic effects of their own or induce the synthesis or release of calorigenic hormones in their target glands. Anterior pituitary extracts were shown to raise metabolic rates markedly by Gaebler in 1933. Among the purified pituitary hormones, growth hormone, the adrenocorticotropic hormone, the lactogenic hormone, the melanophore-stimulating hormone, and some erythropoietically-active fractions have calorigenic effects. The luteinizing hormone was found to be calorigenic, but its effects were accounted for by impurities that had TSH and lactogenic activity. Follicle-stimulating hormone is not calorigenic (Riddle *et al.*, 1936). The thyrotropic hormone is not itself calorigenic. Although TSH regulates mitochondrial respiration in the thyroid gland, it apparently has no direct actions on the mitochondria and is thought to increase ATP consumption indirectly in the stimulated thyroid cell (Lamy *et al.*, 1967).

In vitro, several posterior pituitary hormones make mitochondria swell, but that action seems to depend upon the presence of trace impurities of metal ions; when the metals are removed, no swelling occurs (Cash and Gardy, 1968).

Growth hormone intensifies the action of the thyrotropic hormone on the thyroid gland in hypophysectomized rats, but also intensifies the calorigenic effects of administered thyroxine in such rats; part of the effect of growth hormone is therefore exerted in the peripheral tissues (Evans *et al.*, 1958). The heat production after treatment with growth hormone plus large doses of TSH or T_4 is lethal, unless hydrocortisone is also given. Growth hormone acts synergistically with DNP as well. In the hypophysectomized rats used by Evans *et al.*, 10 µg of DNP per g raised the metabolic rate 23 per cent; that can be compared with a rise of $+26$ per cent in hypothyroid rats and of $+60$ per cent in normal rats in our studies (Hoch, 1965a, b). When Evans *et al.*, treated their hypophysectomized rats with growth hormone for 15 days, DNP raised the metabolic rate $+29$ per cent—a small but definite intensification. Growth hormone raises the metabolic rate in obese women; a dose of 8 mg per day for 8 days produces a 10 to 20 per cent increase in O_2 consumption and a slight rise in plasma free fatty acids on the second 4 days of treatment (Bray, 1969). The lipolysis was thought to cause the increased oxygen consumption.

The mechanism of calorigenesis is obscure. Growth hormone stimulates protein synthesis, and thus might increase heat production either through a general increase in energy transformation incident to amino acid activation and the synthesis of peptide bonds (a sort of specific dynamic action), or through the specific synthesis of excess respiratory enzymes, as suggested by Sokoloff and by Tata for thyroxine calorigenesis.

A purified melanophore-stimulating hormone preparation is reported to be calorigenic (Derrick and Collip, 1953), but the data show considerable scatter. The lactogenic hormone of the pituitary gland is calorigenic in hypophysectomized birds, and also in birds that have been both hypophysectomized and thyroidectomized; when given together with TSH to a bird with a thyroid gland, synergism is seen (Riddle *et al.*, 1936). In rats deprived of their pituitary glands, or both their pituitaries and thyroids, partly purified fractions of pituitary glands that are active in stimulating the formation of red cells in bone marrow are calorigenic also (Contopoulos *et al.*, 1954).

STEROID HORMONES

The adrenocorticotropic hormone (ACTH) exerts its calorigenic effects only on animals that have adrenal glands, and the adrenal *cortical glucosteroids* are apparently the actual calorigenic agents. Although Astwood *et al.* (1953) reported a calorigenic effect of ACTH in adrenalectomized mice maintained on cortisone, their ACTH preparation was only "virtually" free of growth hormone and TSH.

The adrenal glucocorticoids are independently calorigenic and increase the metabolic rates of patients with hypothyroidism (Beierwaltes *et al.*, 1950), for example. Adrenalectomy depresses the metabolic rate of rats about -10 per cent; however, cortisone also has been shown to act in conjunction with the thyroid hormones to control heat production, each hormone raising the metabolic rate by a simple addition of its separate action (Shida and Barker, 1962). Adrenalectomy depresses the metabolic rate of thyroidectomized rats about -20 per cent and cortical extracts restore the metabolic rate to hypothyroid levels (Hoffmann *et al.*, 1948). Such findings suggest corticoids and thyroid act on different target systems.

On the other hand, other studies show a more complex interdependence between cortical steroids and thyroid hormones. Hypophysectomized rats have abnormally low calorigenic responses to administered thyroxine or TSH; ACTH, hydrocortisone, or cortical extracts retore the normal level of response (Evans *et al.*, 1957; Hoffmann *et al.*, 1948). Similarly, the oxygen consumption of leukocytes obtained from patients with secondary hypothyroidism resulting from hypopituitarism is not stimulated by a thyroid hormone (Triac), although leukocytes from primary-hypothyroid patients do respond, and the administration of cortisone to the first group of patients corrects the defect (Murray and Bissett, 1963). A contrary report that Triac or L-T_3 have immediate calorigenic actions *only* in hypophysectomized rats that lack adrenal function and that ACTH abolishes the immediate responses to the thyroid hormones (Donhoffer *et al.*, 1958), is difficult to reconcile with the other studies. Further, studies showing that adrenalectomy increases the activity of rat liver mitochondrial α-glycerophosphate dehydrogenase (Henley, 1962; Henley *et al.*, 1963) would seem to indicate an antagonism between adrenal hormones and thyroid, since thyroid induces the synthesis of large quantities of this enzyme.

It has been thought that such "permissive" effects of adrenal corticoids on the actions of other tropins and hormones might arise from a depletion of the cellular apparatus of either the endocrine gland or the peripheral tissues in corticoid deficiency. Only when corticoids are administered, and enough time passes for resynthesis of polyribosomes and intracellular enzymes, do the cells become fully responsive to tropic or other hormones. A functioning adrenal cortex is necessary for L-thyroxine to stimulate the synthesis of the enzymes involved in carbohydrate metabolism (Freedland *et al.*, 1968), for example. The well-known stimulatory effects of glucocorticoids on protein synthesis in livers might also be a basis for their independent calorigenic actions in a manner similar to that discussed for the growth hormone (above); however, the adrenal corticoids also appear to have actions or effects on mitochondrial function that do not apparently involve the general process of protein synthesis but a more specific synthesis of mitochondria with altered function.

In one of the earliest reports involving a hormonal influence upon protein synthesis in mitochondria (Lowe and Lehninger, 1955), cortisone treatment lowered the degree of polymerization of mitochondrial RNA as judged by the precipitability in 10 per cent sodium chloride; it is still not clear whether this represents a direct hormonal action on the mitochondrion.

Prolonged treatment (6 to 7 days) of subjects, with 1 to 5 mg per day doses of an adrenal cortical steroid (cortisone acetate), results in mitochondria that exhibit uncoupled oxidative phosphorylation (Kerppola, 1960; Kimberg *et al.*, 1968). Under these conditions, the P:O ratio and the respiration in State 3 decrease in liver mitochondria (see also Clark and Pesch, 1956); no measurements of State 4 respiration are reported, so calorigenesis cannot be accounted for from such data. Consistent with uncoupling *in vivo*, the turnover rate of phosphate groups in the ATP of slices of liver or muscle from cortisone-treated animals increases (Derache *et al.*, 1957), as does the activity of a Ca^{2+}-activated ATPase in liver mitochondria (Kimberg *et al.*, 1969). The inhibition of State 3 respiration apparently is exerted at or near all three sites of phosphorylation in the electron-transport chain. The combination of these defects with electron-micrographic data showing a fourfold hypertrophy of the individual liver mitochondria, and with the failure of bovine serum albumin to reverse the respiratory inhibition or the uncoupling (Kimberg *et al.*, 1968), makes it reasonably clear that under the conditions used cortisone induces the synthesis of defective mitochondria, and that effects rather than direct actions of the hormone were observed.

Aldosterone, on the other hand, appears to have a direct action on mitochondrial function. In mice, a single intraperitoneal injection of 2 μg of D-aldosterone acetate per g of body weight lowers P:O ratios in liver mitochondria and slightly stimulates State 3 respiration (Bedrak and Samoiloff, 1966); State 4 respiration was not measured. The changes occur rapidly, about 1 to 3 hours after injection, and are transient, disappearing by 24 hours. The doses used are probably well above the physiologic levels and lead to cell degeneration in the liver. *In vitro*, between 10^{-10} M and 10^{-4} M of aldosterone depresses the P:O ratio and State 3 respiration. In heart muscle, the hormone appears to act on energy metabolism *in vivo* and *in vitro* before it enhances RNA metabolism (Liew and Gornall, 1968). The involvement of aldosterone in the mechanisms of ion translocation in membranes, especially the energy-dependent transport of Na^+ ions (see Edelman and Fimognari, 1968), is interesting to consider in the light of a possible action on mitochondria; another primary site of action appears to be in the cell nucleus.

Not only is there evidence that steroid hormones control the metabolism of mitochondria, but it also appears that mitochondrial metabolism controls the synthesis of steroid hormones from cholesterol in the adrenal cortex. The synthetic processes involve dehydrogenases depending upon NAD^+ and

$NADP^+$, a pyridine nucleotide transhydrogenase, and a mixed-function oxidase which has at its O_2-terminal enzyme a special cytochrome, P450 (see McKerns, 1968).

Like glucocorticoids in the liver, other steroid hormones, among them estrogens and testosterone, stimulate protein synthesis, but no reports of their having a calorigenic effect are on hand. Estradiol apparently counteracts the uncoupling effects of cortisone on mitochondria, but growth hormone, an adrenal ketosteroid, and testosterone do not (Kerppola, 1960); there seems to be a specific interplay between hormone effects via the synthesis of proteins. Women taking oral contraceptives from 1 to 3 years often exhibit striking changes in the shape, size, and appearance of paracrystalline inclusions in mitochondria prepared from liver biopsies, but the changes are not specific (Perez et al., 1969). Sex hormone effects, including stimulation of protein synthesis, tend to be localized to target organs, probably because the injected hormone localizes by binding with specific tissue receptors. Thus, testosterone injection rapidly (1 hour) depresses the ATP content of the prostate glands of castrated rats (Ritter, 1966; Coffey et al., 1968), perhaps owing to a direct action that depresses ATP synthesis or increases ATP utilization. Much later, testosterone and other androgens increase the activity of respiratory enzymes in the genital tissues of such rats (castration has the opposite effect) probably via changes in the rate of protein synthesis.

A definite in vitro action on mitochondria is observed with progesterone, which inhibits respiration like amytal (Chance and Hollunger, 1963a), but that action is not shown to be related to the hormone's biologic effects. Progesterone is thermogenic in humans, cattle, and rats, perhaps via its uncoupling action at high concentrations. In rats, doses of progesterone between 5 and 50 μg per day for 14 days raise body temperature by 0.5°C. Thyroid gland function appears to be increased after injections of progesterone, and in thiouracil-treated rats progesterone is not thermogenic (Freeman et al., 1970). These findings make it appear that progesterone acts by changing the thyroid status; however, because progesterone is thermogenic in thyroidectomized rats, Freeman et al. suggest it may act through the hypothalamo-hypophyseal axis.

12

HYPERMETABOLIC
STATES

DRUG-INDUCED HYPERMETABOLISM

Substituted Phenols

The normal regulatory mechanisms that control energy transformation and expenditure are disturbed in a number of primary and secondary pathologic conditions. The best understood hypermetabolic state is that caused by the administration of chemical agents, of which 2,4-dinitrophenol is the classic example. This group of chemicals acts as uncoupling agents upon isolated mitochondria, and the mechanism of their actions there has been discussed in Chapter 5.

An early report of chemical calorigenesis in living organisms, using nitro-α-naphthol (Cazeneuve and Lepine, 1885), focussed attention on phenolic compounds. The calorigenic action of thyroid gland tissue, observed in 1895 (Magnus-Levy), was connected with the phenolic nature of its active component only after Kendall (1929) described the structure of thyroxine.

DNP was first observed to cause hypermetabolism in humans through investigations of an occupational disease. The events of the period from 1914 to 1918 in Europe evoked the synthesis of large quantities of nitro-substituted phenolic compounds, like trinitrotoluene, for their physical properties. Workers in French munitions plants suffered from a hypermetabolic disorder, with yellow pigmentation, which was often fatal (Magne *et al.*, 1931–32; Perkins, 1919). The disorder was traced to an intermediate compound, DNP.

The early studies on nitrophenols provide a great portion of our present knowledge (see also van Uytvanck, 1931; Heymans and Bouckaert, 1932;

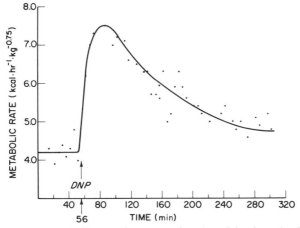

Figure 12–1. The metabolic rate of a rat, as a function of the time after intraperitoneal injection of 2,4-dinitrophenol, 10 μg per g. At zero time, the animal was injected intraperitoneally with diazepam, 40 μg per g, dissolved in a mixture containing water, 15 per cent; propylene glycol, 45 per cent; dimethyl sulfoxide, 30 per cent; ethanol, 10 per cent; this eliminates spontaneous activity for the period of the experiment. DNP was injected at 56 minutes, as indicated. The metabolic rate was measured in the apparatus shown in Figure 7–1; the temperature of the chamber was 20°C. (From Petraitis and Hoch, unpublished results.)

Heymans and Casier, 1933). DNP can be absorbed through the skin or by ingestion and possibly via the respiratory tract. Hyperpyrexia is a striking symptom. Past or present alcoholism sensitizes the subject to DNP. (That phenomenon is reminiscent of the sensitizing effect of ethanol intake on the poisoning seen after carbon tetrachloride inhalation. As noted on page 155, halogenated hydrocarbons can act as uncoupling agents, mainly on liver mitochondria.) *Rigor mortis* appears almost immediately in fatal cases. It was noted that the poison appears to be a general stimulant of the cellular oxidations.

Administration of DNP to animals or humans produces a rapid transient rise in the metabolic rate and in body temperature, if the ambient temperature is above a certain critical temperature, around 23 to 25°C for humans and 20 to 22°C for animals (Shemano and Nickerson, 1959, 1963). Apparently at temperatures low enough, the excess amount of heat produced can be dissipated through compensatory mechanisms set in function by thermoregulatory centers of the hypothalamus; at 2 to 6°, the DNP effects on temperature and metabolic rates are not seen (Tainter, 1934). At higher ambient temperatures, heat production exceeds the capacity of skin radiation, conduction, convection, sweat evaporation, vasodilation, increased cardiac stroke volume, and so forth to drain off heat, and the body temperature rises.

The features of DNP action on the metabolic rate in rats are shown in Figure 12–1. A dose of 10 μg per g given intraperitoneally raises the metabolic rate 77 per cent in about 30 minutes; the rate stays at that level for only a few minutes and begins to fall. Four hours after the injection, the metabolic rate is still elevated about 15 per cent. The effects of various doses of subcutaneously injected DNP on the metabolic rate according to Tainter (1934) appear in Figure 12–2. Equally as striking as the rises in the rate of oxygen consumption are the elevations of body temperature; in dogs curarized to eliminate the contributions of muscle contractions, a subcutaneous dose of 100 μg of DNP per g raises the temperature up to 6 or 7°C and is lethal in about 30 minutes (Magne *et al.*, 1931–32). The latter authors also found that 50 μg per g of DNP dissolved in oil and given subcutaneously killed dogs in 1 to 2 hours.

The calorigenic actions of DNP and its congeners, especially 2,4-dinitro-ortho-cresol, have been recognized as being applicable to yet another human foible: these agents were used to treat obesity. The prospect was that one might continue to overeat without weight gain if one also overburned (Dodds and Robertson, 1933; Dunlop, 1934; Cutting *et al.*, 1933). Unfortunately the safety factor for the nitrophenols is too narrow, and there were fatalities (most occurring during hot weather) traced to their use, besides unpredictable side effects, such as cataracts. The use of these agents as drugs was discontinued and discouraged.

There is a surprising number of ways in which the phenolic uncoupling agents may be accidentally introduced into the human body. The chief compounds found in poisoning cases are dinitro-ortho-cresol, DNP, and pentachlorophenol. They are used for impregnating wood and for spraying insects and weeds, and are usually inhaled, although oral and percutaneous absorption also occur. It is difficult to obtain data on the incidence of such poisonings, but Moeschlin (1965) advises that "in case of poisoning with a high rise in temperature, the possibility of dinitro-ortho-cresol poisoning should therefore always be borne in mind."

The acute actions of sublethal doses of the phenolic uncoupling agents are transient, and after a few hours the subject apparently returns to a normal

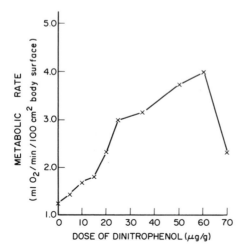

Figure 12–2. The metabolic rate as a function of the dose of dinitrophenol. Rats were injected subcutaneously with DNP, and their metabolic rates measured in a closed system kept at 27°C; the means of the maximal rates, occurring within an hour after injection, are plotted. The metabolic rate is expressed as ml O_2/min/100 cm² of body surface; body surface = 11.34 × (weight)$^{0.67}$ (Replotted from Tainter, 1934.)

metabolic state (Ambrose, 1942). Single doses increase the excretion of creatine, creatinine, and inorganic phosphate. Repeated sublethal doses produce weight loss, depletion of liver glycogen, and a negative Ca balance mainly through increased fecal loss (Pugsley, 1935); the repeated episodes of defective energy transformation and loss of P_i probably account for these observations. Liver damage and hepatic coma are observed in chronic poisonings.

A feature of lethal poisoning with uncoupling agents is the almost immediate onset of *rigor mortis*. In frogs or rats given DNP or dinitro-ortho-cresol, muscle respiration and the anaerobic accumulation of lactic acid are increased up to 10-fold; increased hydrolysis of muscle ATP and decreased synthesis of ATP and phosphocreatine lead to the rapid disappearance of all the muscle ATP and phosphocreatine (Ehrenfest and Ronzoni, 1933; Ronzoni and Ehrenfest, 1936; Parker *et al.*, 1951; Parker, 1954; Stoner *et al.*, 1952). The function of $\sim$P-bonds in muscle contraction seems to be to relax or lengthen the actomyosin polymer fibers; in their extended forms, the fibers are in an energy-rich state and the stimulus for contraction rapidly converts that potential energy into kinetic energy. The rapid onset of *rigor mortis* seems to be a specific sign of mitochondrial damage severe enough to deplete muscle $\sim$P-bonds.

There is at present no well-documented way to stop or reverse the fatal hyperpyrexia after nitrophenol poisoning except perhaps to provide enough cooling of the body to lower the extreme temperature rises. (This is not true for all calorigenic agents. The calorigenic action of thyroid hormones is blocked both *in vivo* and *in vitro* by Mg^{2+} ions and other agents [see Chap. 14]; however, the other phenolic uncoupling agents are not blocked by Mg^{2+}, one of the findings that indicates that calorigenesis proceeds via different mechanisms.) Observations that various agents or drugs antagonize the hypermetabolic effects of phenolic uncoupling agents require more investigation before they are useful therapeutically. The administration of glucose or phlorizin to rabbits previous to the injection of DNP almost completely blocks DNP-induced hyperthermia; insulin strongly augments the DNP-action (Narusawa, 1956). The temperature rise, the phosphaturia, and the decreased acid-labile serum phosphate that occur after rabbits receive DNP intravenously are all reported to return to normal levels after intravenous treatment with quinine (Okui, 1955). Antipyrine or sulpyrine are antipyretic but have no effect on the changes in phosphate metabolism. *In vitro*, quinine counteracts the decreased utilization of P_i that follows the addition of DNP to rabbit kidney homogenates oxidizing succinate.

Hofmann-Credner and Siedek (1949) have reported the prevention of the calorigenic action of dinitro-ortho-cresol by pretreatment with methylthiouracil. In rats, pretreatment or concurrent treatment with methylthiouracil not only prevents the lethal effects of dinitro-ortho-cresol in a chronic treatment program, but also saves all the rats receiving a large dose of dinitro-ortho-cresol that kills control animals in a few hours. Methyl-thiouracil decreases the calorigenic effect of dinitro-ortho-cresol at a time well before methylthiouracil has itself decreased the BMR. Siedek (1950) reports that intravenous administration of 10 ml of 2.5 per cent sodium methyl-thiouracil rapidly lowers the metabolic rate in persons suffering from dinitro-ortho-cresol-induced hypermetabolism, and attributes the antagonism between the two agents to a peripheral effect of methylthiouracil similar to its antagonism to thyroxine. This finding, if confirmed, is relevant to the facilitating effect of the thyroid hormones upon the calorigenic actions of a number of agents (see Chap. 14). Reports that actinomycin D prevents and reverses the calorigenic actions of uncoupling agents, like DNP and methyl-ene blue (Rossini and Salkho, 1966), are also discussed in Chapter 14.

The uncoupling agents exert specific actions on mitochondria, and that specificity has been used experimentally to determine whether certain metabolic processes involve energy-rich phosphate bonds. If DNP inhibits a process, it is usually assumed that the process involves $\sim$P-bonds, and con-versely if DNP does not inhibit, no $\sim$P-bonds are required. Such studies have demonstrated the energy-dependence of peptide-bond formation, fatty acid oxidation, osmoregulation, and photosynthesis, among others. In addition, it is possible to test whether the rate of a particular respiration is controlled by the phosphate cycle; only if it is, should dinitrophenol change the rate (Simon, 1953).

That DNP and other substituted phenols raise the metabolic rate by acting directly on mitochondrial oxidative phosphorylation is well docu-mented experimentally (Dianzani and Scuro, 1956; Dianzani and Dianzani-Mor, 1957; Ernster and Luft, 1964). Studies with injected uncoupling agents have demonstrated not only functional changes but also the presence of the agent in mitochondria: about 60 μg of dinitro-o-cresol or DNP per g of protein in rat muscle or liver mitochondria is found after injecting a dose of 10 μg per g body weight (Buffa *et al.*, 1960, 1396; Carafoli *et al.*, 1963; Hoch, 1968b), and 500 μg of pentachlorophenol per g of protein in rat liver mitochondria after injecting 30 μg per g (Weinbach and Garbus, 1966). The compounds accumulate in mitochondria extremely rapidly, maximal amounts being found when the mitochondria are isolated from animals killed from 3 to 12 minutes after administration of the reagent. The binding of the phenol to the mitochondrial proteins influences the amount found; pentachlorophenol is bound more strongly than DNP and is not as easily removed from the mitochondria by washing. The relative looseness of DNP binding, as well as the transience of the functional changes, is probably the basis for some investigators (Parker, 1956) finding no depression of respira-tory control or P:O ratio in mitochondria from rats injected with DNP.

We have identified DNP in liver mitochondria after injection of rats with DNP-^{14}C, and shown functional changes that are consistent with the presence of DNP (Hoch, 1967b, 1968b). When rats are sacrificed 2 minutes after injection of DNP, their liver mitochondria oxidize glutamate in State

Table 12–1 *Oxidative Phosphorylation in Mitochondria Obtained from Livers of Rats Injected with DNP* (10 $\mu g/g$) *and Sacrificed 2 Minutes Later. Substrate:* (*I*) *Glutamate;* (*II*) *Succinate**

		RESPIRATION (μl O₂/hr/mg protein)			
EXPT	SOURCE OF MITOCHONDRIA	*State* 4	*State* 3	RESPIRATORY CONTROL	P:O
I	Normal rats	13.2	45.6	3.5	2.7
	Normal rats + DNP, 10 $\mu g/g$, 2 min	19.2[a]	45.0	2.3[a]	2.1[a]
II	Normal rats	25.5	57.3	2.3	1.7
	Normal rats + DNP 10 $\mu g/g$, 2 min	30.5	45.5	1.5[a]	1.7

$p <$ [a]0.05

* From Hoch, 1968b.

4 45 per cent faster and succinate 20 per cent faster than those from control rats (Table 12–1). Respiratory control falls significantly with both substrates, and the P:O ratio declines when glutamate is oxidized. The function of liver mitochondria obtained from rats sacrificed between 10 and 30 minutes after injection (not shown) progressively returns toward normal.

The presence of uncoupling agents decreases the degree to which Amytal inhibits mitochondrial respiration (Change and Hollunger, 1963a). In the liver mitochondria from DNP-injected rats, Amytal inhibits respiration significantly less than in mitochondria from controls (Table 12–2), as if there were DNP present.

The amount of DNP that is found in liver mitochondria obtained from DNP-injected rats is more than enough to account for the observed changes in function. Assuming that the water content into which DNP can diffuse in the mitochondria from normal rats is 2.56 ml per g of protein (Klingenberg and Pfaff, 1966), the concentration of DNP if all ¹⁴C represents DNP would have been about 140 μM; assuming the total water content, 6.7 ml per g protein, is available, the concentration would be about 50 μM. These are concentrations well within the ranges effective when DNP is added *in vitro.*

Table 12–2 *Effect of DNP Injection on the* in vitro *Amytal Inhibition of Respiration in Rat-Liver Mitochondria**

	% INHIBITION OF STATE 3 RESPIRATION		
SOURCE OF MITOCHONDRIA	*0.16 mM Amytal*	*0.32 mM Amytal*	*0.48 mM Amytal*
Normal rats	26	54	74
Normal rats + DNP, 10 $\mu g/g$, 2 min	15[a]	38[c]	55[b]

$p <$ [a]0.001; [b]0.005; [c]0.025.

* From Hoch, 1968b.

The mitochondria used in the studies cited here were obtained from the livers of rats. It is not as easy to demonstrate injected DNP in the mitochondria prepared from skeletal muscle, and those mitochondria fail to show functional changes after isolation, even when the rats are injected with very large, lethal doses of DNP. The high rise in metabolic rate, the high respiration of muscle under these conditions, and the appearance of *rigor mortis* make it very likely that the muscle mitochondria were uncoupled; the isolation procedure must have removed the DNP and restored normal function. Pentachlorophenol and dinitro-ortho-cresol, which are bound more firmly than DNP, can be found in skeletal muscle mitochondria after parenteral administration.

There is a disparity between the rapid appearance of the injected uncoupling agent in liver mitochondria (2 minutes), and the slower rise in the metabolic rate, which becomes maximal at about 30 minutes after injection, when the liver mitochondria function normally again and contain no agent. It is possible that the degree of binding decreases with time, but there is no proof, and the disparity still awaits clarification.

Salicylates

The salicylates are substituted phenols, and share some of their actions. The calorigenic action of toxic amounts of salicylates has been known since the turn of the century, when Singer (1901) showed that asiprin raises the metabolic rate of rabbits. Densi and Means (1916) gave salicylate in large doses (6.6 g per day) to two normal men, and one showed an increased metabolic rate, phosphaturia and uricosuria, and a negative nitrogen balance.

Dodd and Minot (1937) in explaining salicylate poisoning showed that in children salicylates promptly increase both the production and the elimination of heat. No major symptoms occur if the water intake is normal and heat elimination is not interfered with. When heat elimination is interfered with and dehydration is allowed to occur, otherwise harmless doses of salicylates can cause death through hyperpyrexia and exhaustion. In dogs anesthetized with sodium barbital and given 100 μg per gram of sodium salicylate intravenously, there is an immediate rise in O_2 consumption. When kept at normal ambient temperatures the dogs die in several hours, but if the dogs are cooled actively, a dose of 500 μg per gram is not fatal. These observations are an example of the generality that the calorigenic actions of uncoupling agents lead to severe hyperpyrexia and death at lower doses when heat elimination is not adequate. Similar observations have been made with DNP and other agents. The therapeutic measures against salicylate poisoning are as nonspecific as those against DNP-poisoning.

Salicylates raise the rate of oxygen consumption in man (Cochran, 1952), in dogs and other animals, in tissues removed from animals injected with the drug (Brody, 1955, 1956), and in isolated tissues (Sproull, 1954). In humans given ordinary therapeutic doses orally, the metabolic rate rises

+30 to 40 per cent, and when enough salicylate is given intravenously to produce plasma concentrations of 20 to 30 mg per cent, the metabolic rate rises +60 to 70 per cent. In patients with fever, 2 g of aspirin every 4 hours raises the metabolic rate or leaves it at a high level, but lowers the body temperature at the same time (Cochran, 1954). The antipyretic effect of therapeutic doses of salicylates is, however, still unexplained. It depends presumably upon anti-inflammatory effects in the tissues, but DNP does not appear to have anti-inflammatory effects similar to those of salicylates, so the production of uncoupling per se does not seem to explain the therapeutic effects of the salicylates. Phenol itself lowers normal temperatures.

Salicylates behave like a typical uncoupling agent. They accelerate respiration at low concentrations and depress respiration to normal levels and below at higher concentrations (Alwall, 1939). Salicylates uncouple when added to isolated mitochondria in concentrations somewhat higher than necessary with DNP (Brody, 1956; Whitehouse, 1964). The calorigenic effects of salicylates in the whole animal thus seem reducible to uncoupling effects on mitochondria, like those of DNP (Falcone, 1959).

That conclusion is probably an oversimplification. Obviously salicylates have antipyretic effects that are not accounted for by their calorigenic actions. When brain or liver slices are incubated with salicylate, DNP, or both, the combination stimulates respiration more than the sums of the independent stimulations. The potentiation indicates to Sproull (1957) that the actions of salicylate and DNP as metabolic stimulants *in vitro* are neither identical nor independent.

An effect of salicylates that raises plasma Mg concentration has been evoked to explain the antipyretic actions (Charnock *et al.*, 1962). At a dose level of salicylate enough to produce a plasma salicylate of about 50 mg per cent, the plasma Mg concentration rises from its control level in rats, 2.9 mg per cent, to 3.7 mg per cent. Since Mg itself reduces body temperature, the antipyretic effects of salicylates might be mediated through Mg^{2+} ions. The combination of Mg salts with salicylate or aspirin does have a much greater antipyretic effect in rabbits than each independently (Winter and Barbour, 1927–28; Climenko, 1936).

The substituted phenols discussed so far are well-studied examples of agents that produce hypermetabolism through their uncoupling actions on mitochondrial oxidative phosphorylation. A large number of other agents produce hypermetabolism or uncoupling, or both, but have little or no structural likeness to the phenols. Table 12–3 lists some of the latter compounds, with no claim for completeness. There is, in addition, another varied group of compounds that have hypermetabolic effects, but have never been shown conclusively to be uncoupling agents; prominent among this group are the catecholamines and the agents that potentiate catecholamine effects (like amphetamine). Uncoupling causes hypermetabolism, but there are more ways to produce hypermetabolism than by way of uncoupling or loose-coupling.

Table 12-3 *A Partial List of Agents that Uncouple Oxidative Phosphorylation When added to Mitochondria, and Some Correlations With in vivo Studies* *

	IN VITRO	IN VIVO			
COMPOUND	Uncouple Mitochondria	Uncouple Mitochondria	Calorigenic or Thermogenic	Thyrotoxic Sensitization	Hypothyroid Desensitization
Phenols					
2,4-Dinitrophenol	+ +	+ +	+ +	+ +	+ +
Dinitro-o-cresol	+ +	+ +	+ +	+ +	+ +
Pentachlorophenol					+
Thymol	+ +	+	+ +	+	
Salicylates	+ +		+	+	
Salicylanilides					
Tetrahydro-β-naphthylamine					
Halogenated Hydrocarbons					
Chloroform	+ + + +		+		
Carbon tetrachloride					
Halothane	+	0			
Fatty acids		+ +	+ +	+ +	
Promazines (Chlorpromazine)					

Agent					
Quinacrines (Atabrine)	+				
Antibiotics (Gramicidin; Tetracyclines; Usnic Acid; Humulone)	+		+		
Redox Dyes (methylene blue; cresyl blue; malachite green; indophenols; tetrazoliums)	++	+			
Barbiturates (Thiopental)	++		+0	++	+
Aryl acetic acids (phenylbutazones; 4-isobutyl-phenylacetic acid)	+				
3,5-diiodo-4-hydroxybenzonitrile (Ioxynil, a herbicide)	+++				
Coumarins (Dicoumarol)	+++		+		
Carbonyl cyanide phenylhydrazone	+++++				
Adrenochrome			+	+	
Benzimidazoles	++++				
Bilirubin					
Phlorobutyro phenones (Desaspidin)	+				
Ca^{2+} ions; P$_i$					
Hormones (Thyroxine; insulin; cortisone, aldosterone; glucagon [?]; catecholamines [?])	+0	+	+		+

+ positive action; 0 absence of action; ? questionable positive data.

* Further details, references, and some agents can be found in Brody (1955), Dianzani and Scuro (1956), Loomis and Lipmann (1948), Weinbach and Garbus (1969), Parker (1965), Wynn and Fore (1965), Simon (1953).

NUTRITIONALLY INDUCED HYPERMETABOLISM

Deficiency of Essential Fatty Acids

Animals fed a diet deficient in fatty acids develop a number of symptoms that include skin lesions, failure to grow, and hypermetabolism. These symptoms are cured by the administration of one of what has come to be called the "essential fatty acids." The essential fatty acids are unsaturated in part, and include the C18-acids linoleic (2 unsaturated bonds) and linolenic (3 unsaturated bonds), and the C20-acid arachidonic (4 unsaturated bonds). The rat, dog, and mouse do not synthesize these polyunsaturated fatty acids. Humans probably require the essential fatty acids in the diet, but apart from skin symptoms in infants fed special diets, it is difficult to demonstrate deficiencies because of the wide distribution of these fats in foods.

The basal metabolic rate of rats on a fat-deficient diet rises in the first few weeks (Wesson and Burr, 1931; Burr and Beber, 1937). Burr and his collaborators wondered about an involvement of the thyroid gland, and recent data indicate that there may indeed be a concomitant hyperthyroid state in fat deficiency, in view of the plasma protein-bound iodine being $+80$ per cent above normal after 18 weeks (Gambal and Quackenbush, 1968). However, it is not yet clear that hyperthyroidism causes the observed hypermetabolism, and a number of other investigators have demonstrated mitochondrial defects that do not seem to be attributable to simply hyperthyroidism.

The mitochondria isolated from the livers of rats deficient in unsaturated fatty acids show defects in oxidative phosphorylation and have an abnormal fat content. The efficiency (Klein and Johnson, 1954) and respiratory control (Ito and Johnson, 1964; Smith and DeLuca, 1964) are depressed as compared with the mitochondria from control animals, the first authors describing uncoupling and the last loose-coupling. There is still controversy over whether the mitochondria function abnormally *in situ* and *in vivo*, but there seems to be no doubt about the mitochondria from fat-deficient rats being more labile and sensitive than normal mitochondria to damage by preparative procedures by exposure to moderate temperatures (30°) and by agents like DNP and digitonin (Johnson, 1963; Ito and Johnson, 1964; Smith and DeLuca, 1964). They seem more resistant than normals to lipid peroxidation and swelling (Stancliff *et al.*, 1969). The mitochondria are swollen when isolated, and swell more rapidly afterward than normally; oxidative phosphorylation deteriorates very rapidly. Increased activity of the enzymes of the electron-transport chain has been ascribed to an "unmasking" of latent activity. Succinic dehydrogenase activity increases in the first 2 weeks of deficiency (Hayashida and Portman, 1963), and increased activity of other dehydrogenases and cytochrome oxidase has been observed after varying periods. No measurements of enzyme activities in State 4 or

State 3 are on record, but the observed decreases in respiratory control and in effectiveness of added DNP suggest that mitochondria from fat-deficient animals are partly in State 3u (see Table 5–1) and so do not control their component enzyme rates. The arguments over the functional state of such mitochondria *in vivo* are reminiscent of those over the mitochondria in thyrotoxicosis, and, indeed, even in DNP-poisoning. Nevertheless, the demonstration of a loss of respiratory control offers a basis for the observed elevations in the metabolic rat in fat deficiency.

The defects in oxidative phosphorylation seem attributable to changes in the composition and structure of the mitochondria. In the first week of a deficient diet, before enzyme activity changes, the mitochondrial content of linoleic and arachidonic acids decreases, and eicosatrienoic acid appears (Hayashida and Portman, 1963). In the first 6 weeks the content of essential fatty acids is lowered by 80 per cent, and neutral lipid content doubles. Alterations in the structure have been thought to accompany the changes in mitochondrial lipids, and bizarre forms have been seen by electron microscopy both *in situ* and after isolation of the mitochondria. The lability of the mitochondria has been attributed to such changes, presumably because the membrane lipids are faulty. The mitochondria are apparently constructed with incompetent building blocks in the membranous portions. The lipids of the membranes are known to turn over during the 10-day half-life of the mitochondrion, and so faulty components might be introduced into preexisting mitochondria.

Magnesium Deficiency

Rats that are made deficient in Mg have elevated metabolic rates, and mitochondria obtained from their livers are uncoupled. It is difficult to determine whether the uncoupling is correlated with a low amount of Mg in the mitochondria or is due to an antagonism between Mg^{2+} and thyroid hormones (Vitale *et al.*, 1957a, b). A Mg-deficient diet is reported to increase the size of the thyroid gland in rats and the ^{131}I uptake 24 hours after injection, but there is no rise in the plasma protein-bound I level (Corradino and Parker, 1962). Conflicting data exist; a normal mitochondrial content of Mg has been reported in Mg-deficient animals, and only mitochondrial swelling, with no uncoupling or potentiation of thyroxine-uncoupling (Nakamura *et al.*, 1961; Beechey *et al.*, 1961; Kalant and Clamen, 1959). These discrepancies may represent differences in techniques; it is inviting to think that a deficiency of Mg^{2+} ions in mitochondria should cause hypermetabolism, because *in vitro* an Mg^{2+}-free medium produces mitochondrial swelling and a loss of respiratory control with normal phosphorylation (loose-coupling), and addition of Mg^{2+} salts restores respiratory control. Still, the intellectual pleasure of accounting for the *in vivo* hypermetabolism by a subcellular process probably ought not to outweigh the lack of firm data.

GENETICALLY INDUCED HYPERMETABOLISM

In a single case, severe, continued hypermetabolism has been shown to be connected with the presence of functionally and morphologically defective skeletal muscle mitochondria (Ernster *et al.*, 1959; Luft *et al.*, 1962; Ernster and Luft, 1963, 1964). This Swedish lady, first studied at the age of 35 years, perspired profusely, drank excessive amounts of water, and became progressively thinner and weaker, from the age of 7. Her BMR was $+150$ to $+200$ per cent above normal levels, a degree of elevation not seen in thyrotoxicosis. No exogenous cause for her hypermetabolism could be found, and she did not appear to be thyrotoxic by clinical criteria. Neither a subtotal thyroidectomy nor administration of iodine and thiouracil relieved her hypermetabolism.

Abnormal mitochondria were found in biopsies of her muscles. Electron micrographs revealed morphologic abnormalities, such as strange shapes and peculiar inclusions, and an increased number of mitochondria in the subsarcolemmal spaces. On isolation of the mitochondria, their total protein and cytochrome activity was three to four times above normal levels, and they performed oxidative phosphorylation in a loosely coupled manner. ADP and P_i did not stimulate respiration, but P:O ratios were normal. The respiration (which was presumably in State 3 and so phosphorylating respiration) was not inhibited by oligomycin for some unexplained reason. Activity of the mitochondrial Mg^{2+}-sensitive ATPase was increased, and addition of DNP did not accelerate it—and the respiratory rate—more than very slightly.

It is not known if this lady synthesizes abnormal mitochondria (although the RNA content of her muscle was abnormally high), or if other inborn errors of metabolism kept her mitochondria in a loose-coupled state. No genetic linkage of her condition was mentioned, but it can be assumed that she had a genetically determined defect in her mitochondria.

There are other reports of genetically linked defective skeletal muscle mitochondria that have been obtained from patients suffering from muscle disease but without hypermetabolism. In two unrelated 8-year-old children, distinct clinical and cytologic features have been given the names *megaconial* myopathy and *pleoconial* myopathy (Shy and Gonatas, 1964; Shy *et al.*, 1966). Both had proximal muscle weakness from early infancy, but the latter was distiguished by episodes of severe quadriplegia. The BMRs and the chemical indices for thyroxine metabolism were normal. Examination of muscle biopsies revealed abnormalities in the mitochondria that were different from those seen in Luft's patient. The megaconial myopathy was so called because of the presence of huge mitochondria with few *cristae* that occupied 20 to 40 per cent of the muscle cell, and showed cytochemical evidence of increased activity of pyridine nucleotide and succinate dehydrogenases. The pleoconial picture featured variety in mitochondrial shapes.

Both sorts of mitochondria contained unusual inclusions, some round and some rectangular arrays that are almost crystalline in their regularity (they call to mind electron micrographs of viruses and make one think of the intra-mitochondrial nucleic acids). A few siblings of each child also showed evidence of abnormalities in skeletal muscle function, and it seems clear that a genetic defect exists in these families.

Congestive cardiac failure or congenital myopathies in animals also may produce myocardial mitochondria with decreased oxidative phosphorylation (Opie *et al.*, 1964; Gertler, 1961; Schwartz and Lee, 1962) but other workers report no mitochondrial abnormalities (Blanchaer and Wrogemann, 1968).

Some of the considerations in correlating examinations of isolated skeletal muscle mitochondria with physiologic phenomena have been discussed here (see p. 113) in connection with thyrotoxic myopathy. Findings of loosely coupled oxidative phosphorylation in other human myopathies (van Wijngaarden *et al.*, 1967; Hulsmann *et al.*, 1967) are also evaluated by Hulsmann *et al.* (1967). Part of the difficulty in interpretation seems to arise from the techniques of preparing mitochondria from skeletal muscles. It is hard to understand how a loss of respiratory control in skeletal muscle mitochondria would not be accompanied by an elevation in the basal metabolic rate—why are the children with myopathy and normal BMRs not like the hypermetabolic Swedish lady?

General References: Early work on DNP is covered by Simon (1953). The relation between the BMR and mitochondrial respiratory control and the Swedish lady are discussed by Ernster and Luft (1964). M. J. H. Smith (1963) reviews the metabolic effects of the salicylates.

CHAPTER

13

SENSITIZATION TO CALORIGENIC ACTIONS

The calorigenic actions of administered agents depend on the physiologic state of the subject. Probably the earliest recognized example of this phenomenon is the dependence of the effects of administered epinephrine on the thyroid state. Hyperthyroidism is accompanied by an increased calorigenic action of many, if not all, uncoupling agents of exogenous nature, and some of endogenous nature. A biochemical rationale for this generality is presented in Chapter 8. The experimental testing of such a model has more than just theoretical importance.

The phenomenon of potentiation, or perhaps better, *synergism*, refers to the effects of combinations of agents. Two drugs may have combined effects equal to the sum of their individual effects (summation), or a supra-additive effect; one of the pair may have no effect itself, but may cause the other to have an increased action. Goodman and Gilman (1967) prefer to reserve the term synergism strictly for heterergic drugs when the combined effect is greater than that of the active component alone. Synergism applies strictly to a supra-additive action of an administered calorigenic agent when the subject has been *treated* with thyroid hormones first. Actually, a spontaneously or "endogenously" hyperthyroid individual would be expected to exhibit the same phenomenon of hyperaction to a dose of an administered uncoupling agent. The term *sensitization* is used here phenomenologically to refer to the augmented effects of administered agents, whatever the mechanism. The opposite of this augmentation would be "desensitization," although that term has been preempted by the immunologists. Excess thyroid hormone sensitizes toward many uncoupling agents, and deficiency of thyroid hormone desensitizes.

SUBSTITUTED PHENOLS AND THYROID HORMONES

Calorigenic sensitizations were first observed through measurement of temperature. Borchardt (1928) showed that thyroidectomized or adrenalectomized cats respond slightly less than normally (are desensitized) to the febrile stimulus of subcutaneously administered tetrahydronaphthylamine. Animals deprived of both thyroid and adrenal glands have no febrile responses to tetrahydronaphthylamine, but when they are pregnant they respond normally (via the hormones produced by the fetus). Treatment of thyroidectomized or adrenalectomized cats with small doses of thyroid hormone or an adrenal preparation restores the normal response. In normal rabbits treated first with thyroxine, a dose of tetrahydronaphthylamine that produces only a slight fever in normal animals produces a fever of 44.5°C and is rapidly lethal (Glaubach and Pick, 1930, 1931). The calorigenic actions of cocaine, procaine, and novocaine are also made more severe by pretreatment with thyroxine. A dose of dinitrophenol that raises the body temperature of normal rabbits only slightly is lethal in a few hours in hormone-pretreated rabbits (Glaubach and Pick, 1934). Glaubach and Pick suggested caution in the administration of other drugs at the same time as thyroid hormone.

Their warning was well taken. Poole and Haining (1934) reported a sudden death of a patient from DNP poisoning. An obese woman who had been taking desiccated thyroid, 50 mg per day, for 18 months became comatose a few days after starting to take DNP. She had a high fever and flaccid muscles and died soon after admission. Autopsy showed intercellular liver edema and fragmentation of the myocardium. Poole and Haining concluded that "there is a possibility on which we are not competent to pass judgment, that the fatality was due to some vicious synergism between the action of dinitrophenol and of thyroid." This fatal outcome may have depended upon dosage or perhaps on yet other sensitizing agents besides thyroid, for Simkins (1937) treated 181 obese patients with DNP and thyroid, and all lived and lost weight, more than with DNP alone; the worst feature of the treatment was the incidence of cataracts.

Sensitization by thyroid hormones is demonstrable when the respiration of tissues rather than of the intact subject is measured. The perfused thighs of dogs respire +50 per cent to +124 per cent faster when DNP is added to the perfusate; feeding fresh thyroid to the donor dogs before perfusion doubles the effect of DNP (Alwall, 1936; Alwall and Scheff-Pfeiffer, 1936). Methylene blue, another uncoupling agent, acts like DNP in this system;

however, although perfusing with thyroxine accelerates respiration, it does not sensitize this tissue respiration toward the uncoupling agents.

Measurements of the metabolic rate confirm that administered thyroid hormone sensitizes normal animals toward calorigenic agents, e.g., in rats treated with thyroxine and dinitro-ortho-cresol (Barker, 1946), or T_3 and sodium salicylate (Gemmill *et al.*, 1962). Hypothyroidism desensitizes as shown by the minimal effects of DNP or salicylate in rabbits, rats, and pigeons (Alwall, 1936; Gemmill *et al.* 1962; Riddle and Smith, 1935).

These sensitizing effects of thyroid hormones can apparently be blocked by certain agents that act as inhibitors of respiration or of protein synthesis (see Chap. 14). When a tissue becomes more functionally active, the metabolism, especially oxidative and phosphorylative, usually increases, but the character of the metabolism is generally modified so that the inhibitor-sensitive and inhibitor-resistant portions do not change equally. In rabbits sensitized toward DNP by pretreatment with thyroxine, urethane anesthesia is reported to abolish the sensitization selectively, without affecting the normal degree of DNP-induced calorigenesis (Alwall and Sylvan, 1937, 1939) (Table 13–1).

Table 13–1 *Metabolic Rates ($ml\ O_2/kg/10\ m$) in Rabbits After Receiving 15 µg of DNP/g Subcutaneously**

ANIMALS	— URETHANE	+ URETHANE
Normal	53	59
T_4-treated	163	60

* From Alwall and Sylvan, 1937.

Urethane is known to inhibit electron-transport, like other narcotics, by blocking the oxidation of cytochrome b and the reduction of cytochromes, c and $a + a_3$ (Keilin, 1925; Keilin and Hartree 1939). In tissue homogenates, a low concentration of urethane inhibits five times more strongly the elevated respiration caused by adding DNP than it does the unstimulated respiration (Florijn *et al.*, 1950).

Alwall and Sylvan postulated that urethane blocked the peripheral action of thyroxine. Their working hypothesis can now be translated into up-to-date terms. The "dehydrogenation gradient," i.e., the difference between the capacity of tissues for dehydrogenation (State 4) and for H-transfer (State 3), represents the limits of the degree to which DNP can stimulate respiration. Thyroxine raises the dehydrogenation capacity (State 4), but perhaps secondarily to preoxidative processes (protein synthesis, perhaps), without raising the H-transfer capacity (State 3); the H gradient is thus increased, and DNP can stimulate respiration more than normally. Deep urethane narcosis was supposed to make even thyroid-treated animals functionally athyroid, and so abolish the synergism; whether that has any verisimilitude or not remains to be seen, but it is one of the few examples of a therapeutic attempt against the manifestations of the thyroid-uncoupler synergism.

Recent studies have shown that the *in vivo* phenomenon of thyroid hormone sensitization is accompanied by altered *in vitro* sensitivities of mitochondria to the uncoupling agents, which seem to account for the *in vivo* phenomena. These studies, and a rationale for sensitization, are discussed in the chapter on the mechanisms whereby thyroxine controls energy transformations (Chap. 8).

HALOGENATED HYDROCARBONS

Rats pretreated with doses of thyroxine not large enough to cause them to lose weight, and then treated with a subcutaneous dose of chloroform ($CHCl_3$) in oil, which when given to untreated rats is anesthetic but not lethal, die within 48 hours (McIver, 1940; McIver and Winter, 1942). In one rat, the temperature was 104.6° at the time of death, but no mention was made about the others. The low amount of glycogen in the liver is not a major cause of the increased sensitivity to $CHCl_3$, and a high-protein diet gives no protection (as it might be expected to if simple hepatic necrosis were the cause of death). McIver (1940) speculated as to the possible application of these findings to the choice of anesthetic for operations upon patients with hyperthyroidism, because other anesthetics besides $CHCl_3$ were capable of causing liver damage, and so their effects as well as those of $CHCl_3$ might be intensified (see Chap. 16).

Carbon tetrachloride, CCl_4, also becomes more lethal after thyroid treatment. Calvert and Brody (1961) pretreated rats with 0.5 mg of T_4 subcutaneously every day for 5 days, and 5 days after the last treatment gave CCl_4 by stomach tube. Neither treatment alone was fatal, but combined treatment killed the animals within 20 hours. Liver mitochondria obtained from animals killed between 8 and 10 hours after receiving CCl_4 show depressed oxidative phosphorylation *in vitro*. In thyroxine-pretreated rats, CCl_4 lowers the rate of oxidation of glutamate, but less so when NAD^+ is added; the P:O ratio behaves similarly, and there is a high activity of the Mg^{2+}-activated ATPase. Thyroidectomized animals are partly protected against CCl_4 poisoning. Mitochondria from animals thyroidectomized before receiving CCl_4 show lesser degrees of change in oxidative phosphorylation than mitochondria from normal animals.

Carbon tetrachloride and chloroform are uncoupling agents (Table 12–3), especially in the liver where they accumulate. Carbon tetrachloride *in vivo* or *in vitro* causes liver mitochondria to be uncoupled (Dianzani, 1954) and *in vivo* to contain excess Ca (Thiers *et al.*, 1960). Mitochondria prepared from CCl_4-poisoned rats in the presence of chelating agents contain no excess Ca, perhaps because the accumulation of Ca is an artifact of the preparation (Cohn *et al.*, 1968), or because the accumulated Ca is removed by the reagents. The observed accumulation of mitochondrial calcium seems unique to ingested CCl_4, but not to the action of uncoupling agents (Carafoli, 1967; Carafoli and Tiozzo, 1968). In rats injected with DNP or pentachlorophenol, only half the amount of administered $^{45}Ca^{2+}$ is found in liver mitochondria, as compared to normal rats, and the total liver Ca^{2+} is one-third of normal. The latter results seem understandable as an inhibition of energy-dependent Ca^{2+}-uptake by mitochondria; the accumulation of Ca^{2+} in mitochondria after CCl_4 ingestion is not explained by uncoupling.

AMPHETAMINE

It is standard pharmacologic practice to use amphetamine as a "metabolic stimulant" for various purposes. Amphetamine is a calorigenic agent, and raises body temperature and metabolic rate. A dose of 20 μg per g given subcutaneously to rats raises the body temperature by +2.1°C within an hour, and maintains the fever for up to 6 to 9 hours. In similar rats pretreated with thyroxine, amphetamine injection raises the temperature +3.5°C, and half of them die. Urethane counteracts these effects of amphetamine alone or after T_4-pretreatment. Simonyi and Szentgyörgyi (1949) conclude that the hormone acts on the heat center in the nervous system, presumably because that is where amphetamine is supposed to act. The potentiation of amphetamine calorigenesis and toxicity in mice is observed after treatment with T_4 or T_3, and α-methyl-m-tyrosine and ephedrine act like amphetamine, especially under conditions of aggregation (Askew, 1962; Moore, 1966).

The BMR of rats receiving 2 μg of amphetamine per g subcutaneously increases very rapidly and reaches a maximum within 30 to 60 minutes that lasts between 4 and 5 hours (Gyermek, 1950). Pretreatment of the animals with thyroxin makes amphetamine much more calorigenic, and kills the animals rapidly; a dose of 0.5 μg of amphetamine per g raises the metabolic rate four times more in the rats that receive thyroxine for 3 days, and seven to eight times more after 5 days of thyroxine. Gyermek proposes that the synergism is via the sympathetic nervous system.

A patient that ingested dexedrin and desiccated thyroid had a temperature of 104° and a BMR +63 per cent above normal; treatment (successful) was with digitoxin and propylthiouracil given orally over a period of days (Atkinson, 1954).

Hypothyroid animals react subnormally to amphetamine (Simonyi and Szentgyörgyi, 1949; Chu *et al.*, 1969). A dose of amphetamine that increases body temperature and motor activity in normal rats has no such effects in rats made hypothyroid by feeding with methylthiouracil, and the lethal dose of amphetamine is seven times higher (Mantegazza and Riva,

1965). The calorigenic effect of amphetamine is restored relatively rapidly by treating hypothyroid rats with T_3; 0.4 μg per g per day acts in 4 days (Chu et al., 1969).

Amphetamine seems to act as a monoamine oxidase inhibitor and thus a potentiator of the action of the catecholamines. The calorigenic action of amphetamine may be due to the increased effect of catecholamines (Gessa and Clay, 1969). The mechanism of the synergism between amphetamine and thyroxine might be reduced to that of the synergism between catechol-amines and thyroxine; however, the hyperthermia after amphetamine injection seems to arise from events in the diencephalon and seems not to depend upon the sympathetic nervous system, because treatment with ganglion-blocking agents does not moderate the hyperthermia (Belenky and Vitolina, 1962).

Whether amphetamine acts to produce hyperthermia centrally or peripherally, the amphetamine-thyroxine synergism may be very important clinically. Either agent has been used separately for years in the treatment of obesity. Recently, amphetamine and thyroid have been combined with digitalis and other drugs in some proprietary preparations for the treatment of obesity. An occasionally fatal effect of such multicomponent preparations may well be attributed to a synergism of the sort well documented for un-coupling agents and thyroxine.

OTHER SENSITIZATIONS

Chloropromazine, although not thought of usually as a calorigenic agent, causes significant hyperthermia in 10 to 15-day old mice, but causes hypo-thermia in 38-day old mice (Bagdon and Mann, 1965), and raises the BMR of adult rats (Hoch, unpublished data). Chloropromazine uncouples oxidative phosphorylation *in vitro*, and perhaps this is the basis of its calorigenic action rather than its better-known depressant effects on the central nervous system. Animals pretreated with T_4 or DNP become extremely hyperthermic ($>42°$C) and die when given 10 μg per g of chloropromazine. They can be protected to some extent by cooling (Skobba and Miya, 1969).

Phosphate ions, when infused into dogs previously treated with a rela-tively small dose of triiodothyronine, produce severe hyperpyrexia and hypermetabolism, and are fatal in 3 hours (Roberts et al., 1956). Intra-venous administration of phosphates alone does not raise the BMR, nor does the dose of T_3 that Roberts et al. used during the $3\frac{1}{2}$ hours they observed the animals. These striking changes were noted within 60 to 100 minutes after T_3 was given, and so probably depend upon the presence of T_3 in mitochondria. Although phosphates are not calorigenic in these experiments, they do uncouple mitochondria *in vitro* (Hunter and Ford, 1955); therefore, a synergism may exist between phosphate ions and T_3. Another possible mechanism is suggested by Roberts et al.'s observations that phosphate infusion lowered the plasma magnesium ion concentration, presumably through the formation of salts; free Mg^{2+} ions antagonize thyroid hormone actions, and the sudden influx of phosphate ions may remove this restraint upon hormonal uncoupling. These studies suggest caution in administering thyroid hormones to patients with elevated plasma phosphate levels, as in renal disease or hypoparathyroidism.

Dicumarol in doses of 500 to 700 μg per g given intraperitoneally to rats or rabbits causes a severe and lethal hyperpyrexia (110 to 112°F) similar to the effects of dinitrophenol (Seager and Bernstorf, 1951). Dicumarol uncouples mitochondrial oxidative phosphorylation. The *in vivo* pyrogenic effects of this compound are diminished or obliterated in animals made hypothyroid by surgery or propylthiouracil treatment, or by both of these methods. There are no reports on whether this desensitization has the expected counterpart of a sensitization of thyroid-treated animals toward Dicumarol.

CHAPTER

14

HYPOMETABOLIC STATES: ANTICALORIGENIC AGENTS

The hypometabolic states have been referred to here a number of times. The main one is hypothyroidism (Chaps. 8 and 9), but adrenal and pituitary deficiencies also cause low metabolic rates (Chap. 11). A number of agents and agencies also cause hypometabolism; some by acting to counter the effects of thyroid hormones and produce a *de facto* hypothyroidism, others by acting on mitochondria directly.

PERIPHERAL ANTAGONISTS OF THYROID HORMONES

A number of compounds antagonize the actions and effects of the thyroid hormones. Many of them inhibit one or more of the enzyme-catalyzed steps in the thyroid gland to block the synthesis of the hormone. They are used therapeutically to depress to normal levels hormone production when it is excessive. One of them, ^{131}I, eventually destroys all thyroid tissue and makes the subject hypothyroid; it is now the major cause of hypothyroidism in most civilized countries. In the sense of acting on the thyroid gland, these compounds (I^-, CNS^-, ClO_4^-, thiourylidenes, and so forth; see Means *et al.*, 1963) are "antithyroid." In making the subject hypothyroid, they are anticalorigenic agents.

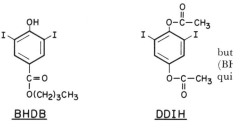

Figure 14–1. The structure of n-butyl - 4 - hydroxy - 3,5 - diiodobenzoate (BHDB) and diacetyl-2,6-diiodohydro-quinone (DDIH).

Another group of agents is "antithyroid" or "antithyroxine" in the sense of blocking the peripheral calorigenic and metabolic effects of the thyroid hormones. They have been developed mainly for therapeutic purposes, because in some situations it would be advantageous not only to stop the synthesis of the hormone, which takes weeks and months to clear up the symptoms of thyrotoxicosis, but also to alleviate the symptoms more promptly. Most of these "peripheral antithyroid" agents are still experimental therapeutically, and are of interest as aids in studying hormone action.

Some peripheral antagonists appear to act because their structure resembles that of the thyroid hormones, but they have no calorigenic effects and compete with L-T_4 and L-T_3 at various sites. The thyroxine-analog 3,5-diiodothyroacetic acid decreases the BMR and suppresses the signs of thyrotoxicosis in human patients when given 1–3 weeks (Frawley and Zacharewicz, 1962). The "reverse" analog of L-T_3 (3,5,3'-triiodothyronine, see Fig. 8–1), which is 3,3',5'-triiodothyronine, and the 3,3'-diiodothyronine act similarly (Pittman and Barker, 1959; Barker *et al.*, 1960; Rawson, 1964). Among the antithyroid compounds that resemble thyroid hormones in structure are n-butyl-4-hydroxy-3,5-diiodobenzoate (BHDB) (Sheahan *et al.*, 1951) and diacetyl-2,6-diiodohydroquinone (DDIH) (Serif and Seymour, 1961). Their structure is shown in Figure 14–1.

Both BHDB and DDIH have a *restricted anticalorigenic effect.* They lower but do not abolish the calorigenic effect of administered L-T_4, but not that of administered L-T_3. Their anticalorigenic effect is less when the dose of L-T_4 is increased, apparently a competitive phenomenon. Because of such findings it has been proposed that they act by inhibiting the deiodination of L-T_4 in the peripheral cells to L-T_3, and that they block L-T_4 calorigenesis because L-T_4 is calorigenic only when converted to L-T_3. The last assumption, however, is not commonly accepted and both L-T_4 and L-T_3 appear to have calorigenic effects. The anticalorigenic effects of BHDB and DDIH are limited to administered L-T_4, and they have no significant effect on the BMR of a normal euthyroid subject. While that may be of therapeutic use it is difficult to visualize how a competition between the drug and the hormone could fail to block the direct action of the hormone, unless the hormone is in a form different from L-T_4.

BHDB acts as an uncoupling agent on mitochondria *in vitro*. It might act on mitochondria *in vivo* as well, and compete there with thyroxine, but how that contributes to its anticalorigenic effect is not clear. It may be pertinent that salicylates also can uncouple *in vitro*, but have an antipyretic effect *in vivo*. BHDB has no antipyretic effect, and in toxic doses produces a fever like the salicylates. BHDB is not only an "antithyroxine" compound; it also has many effects on the metabolism of thyroxine, promoting hormone inactivation by conjugation (Flock and Bollman, 1964). Clinically BHDB seems no better than I⁻ for alleviating thyrotoxicosis, perhaps because it is toxic in doses high enough to be therapeutically effective (Fraser and Maclagan, 1953).

Another peripheral antagonist is thiouracil. Thiouracil and its derivatives are primarily thought of a inhibitors of the iodination of thyroxine in the thyroid gland, and are widely used for treatment of thyrotoxicosis because of that action; however, it has been known for some time that the thiouracils also have a peripheral antithyroid effect (Dietrich and Beutner, 1944; Abelin, 1947; Andik *et al.*, 1949a, 1949b, Barker *et al.*, 1949). When animals or humans are made nypothyroid by surgery or with ^{131}I, and are then given thiouracil, more thyroid hormone is needed to restore euthyroidism than when thiouracil is omitted. Since there is no thyroid gland or endogenous hormone production, that effect of thiouracil has been thought to be peripheral. Thiouracil has a selective anticalorigenic effect that depends on the structure of the iodinated thyronine given: $L-T_4$ has only 7 per cent of the calorigenic potency it has in normal rats receiving thiouracil for up to 9 weeks, whereas the 3,5,3'-triiodothyroacetic and propionic acids are fully potent, and the corresponding tetraiodo derivatives are 30 to 40 per cent potent (Stassilli *et al.*, 1960). Thiouracil inhibits peripheral deiodination of $L-T_4$, like BHDB, and if $L-T_4$ must be deiodinated to become effective, this may be a basis for the antithyroxine action.

The thiouracils and thiourea alter the function of tissues or homogenates *in vitro*. Rather high concentrations, 0.1 to 1 mM, of thiouracil inhibit the responses of heart, gut, or uterine muscle to epinephrine (Raskova, 1948; Friedenwald and Buschke, 1943). Thiouracil and thiourea depress the respiration of homogenates, but thiourea does not block the acceleration of respiration seen when $L-T_4$ is added (Reid and Kossa, 1954), although methylthiouracil is reported to block the action of dinitro-ortho-cresol (Hofmann-Credner and Siedek, 1949; Locker *et al.*, 1950).

It is possible that the thiouracils block the peripheral effects of thyroid hormones by blocking protein synthesis (see below). The incorporation of these substituted uracils into RNA to produce a nonsense RNA that does not code for protein synthesis is demonstrated in tadpole livers (Paik and Cohen, 1961) but not in rat livers (Lindsay *et al.*, 1965), although protein synthesis seems to be slowed in the livers of thiouracil-treated rats (Yatvin *et al.*, 1964). The thiouracils are reported to compete with uracil and thereby inhibit uridine phosphorylase, but they are not substrates (Lindsay *et al.*, 1968); in tissues where uridine phosphorylase activity is rate-controlling, such inhibition may depress nucleotide and RNA synthesis.

Another group of compounds, which has a similar restricted anticalorigenic effect on thyroxine-induced hypermetabolism, shares the property of *inhibiting protein synthesis* at various steps. These agents are anticalorigenic rapidly after administration, and also over periods of hours. Puromycin (20 μg injected intraperitoneally in 2 doses 45 minutes apart) *acutely* (within 1 hour) lowers the metabolic rate far more effectively in hyperthyroid rats than in normal rats, and restores the metabolic rate of the thyrotoxic animals to normal, indicating that a larger fraction of the total body O_2 consumption is "related" or coupled to protein synthesis in hyperthyroidism than in the normal state, and that the increased metabolic rate in hyperthyroidism is secondary to the effect of the thyroid hormones on protein biosynthesis (Weiss and Sokoloff, 1963).

One should differentiate, in considering results with inhibitors like puromycin, between protein biosynthesis as a *process* that uses oxidative energy (and is calorigenic by a sort of specific dynamic effect), and as an *end result*, the synthesis of more oxidative enzymes. Either of these will raise the metabolic rate. Sokoloff's data indicate that it must be only the process under his experimental conditions—how could puromycin lower the BMR within 1 hour if the elevation depended upon more respiratory enzymes being present? The experimental conditions used by Tata are different; they injected hypothyroid rats with L-T$_3$ and an inhibitor of protein synthesis once at zero time and again at 48 hours, and then measured the BMR at 72 hours (Tata, 1963; Widnell and Tata, 1966). The injection of puromycin, actinomycin D, 5-fluorouracil, or cycloheximide prevented the calorigenic effect of the hormone. Under these conditions it seems likely that the end result of protein synthesis was absent, not the process per se at the time of measuring the BMR; an excess of respiratory assemblies was not synthesized.

These studies indicate that the calorigenic effects of thyroid hormones depend upon both the general process of protein synthesis and the specific synthesis of respiratory enzymes, if one accepts the assumption that the agents that block calorigenesis act *only* to block protein synthesis. Certainly there is good evidence to indicate that the thyroid hormone does induce the *de novo* synthesis of respiratory and other enzymes, and that inhibitors of protein synthesis block the induction. A particularly well-studied system is the carbamyl phosphate synthetase that is induced by thryoxine in the livers of tadpoles, an early biochemical change designed for the new pathways of nitrogen metabolism as metamorphosis proceeds. Actinomycin D or puromycin inhibits the increase in enzyme activity when administered together with L-T$_4$ or up to 12 hours after; after enzyme induction has started, the inhibitors do not block further increases (Kim and Cohen, 1968). The findings indicate that once the proper messenger RNA has been formed it can support further enzyme synthesis. Similar data on mammalian systems are not yet available, and there the problem of measuring *de novo* synthesis of respiratory enzymes is much more difficult than in tadpole livers, which have very low levels of carbamyl phosphate synthetase.

However, there is increasing evidence that inhibitors of protein synthesis have effects on respiratory systems that may not be related at all to protein synthesis (Revel et al., 1964). Actinomycin D inhibits the respiration of human leukemic leukocytes and decreases their ATP content; puromycin does not inhibit their respiration (Laszlo et al., 1966). Puromycin, given to mice in doses that inhibit protein synthesis, produces swelling of neuronal mitochondria, although another inhibitor, acetoxycycloheximide, produces no swelling (Gambetti et al., 1968). Puromycin inhibits the respiration of guinea pig cerebral cortex slices, but cycloheximide does not; the former changes mitochondrial morphology (Jones and Banks, 1969). Chloramphenicol inhibits the synthesis of mitochondrial enzymes specifically at low concentrations but may directly inhibit cell respiration at higher concentrations (Firkin and Linnane, 1969). In plants, cycloheximide accelerates respiration and depresses protein synthesis much like DNP, but is more potent (MacDonald and Ellis, 1969). Rossini and Salkho (1966) report that not only does actinomycin D injected together with L-T_3 or L-T_4 block the rise in metabolic rate between 36 and 96 hours later, but also that actinomycin D injected 1 hour before, simultaneously, or 1 hour after the uncoupling agents DNP or methylene blue, blocks or very rapidly reverses the rise in BMR. These uncoupling agents act directly and rapidly on mitochondria, and protein synthesis has never been invoked to explain their actions. (The author has not been able to repeat these experiments with DNP on normal rats.)

Injected actinomycin D very rapidly (30 minutes) reaches the livers of rats (Marchis-Mouren and Cozzone, 1967), and while most of it is found in the cell nuclei, a significant portion, about 10 per cent is found in the mitochondria (Dingman and Sporn, 1965). A portion of the very rapid anticalorigenic effect of actinomycin D may be exerted in the mitochondrion by antagonizing the action of the thyroid hormones. A complex may form between actinomycin D and thyroxine, as shown by changes in the absorption spectrum of the former, and by the ability of a large excess of the hormone to prevent actinomycin D from inhibiting the growth of *Bacillus subtilis* (Kim et al., 1967). Kim et al., suggested that the failure of cultured kidney cells to take up thyroxine in the presence of actinomycin D (Siegel and Tobias, 1966) may be due to the failure of the complex to penetrate membranes. The slower antagonism might involve the observed inhibition of the release of thyroxine from the thyroid gland (Halmi et al., 1969). On the other hand, there is as yet no direct evidence that actinomycin D or the other inhibitors of protein synthesis acts on mitochondrial oxidations directly. Treating rats with the aminonucleoside of puromycin, which produces the nephrotic syndrome, also depresses State 3 respiration in mitochondria obtained from their kidneys but does not depress respiratory control in their liver mitochondria; nor does its addition to isolated normal kidney mito-chondria alter respiratory function (Johnston and Bartlett, 1965). Addition of actinomycin D to lamb heart mitochondria strongly inhibits the incor-poration of ATP into MRNA (Kalf, 1964), but there is no evidence that such incorporation involves oxidative phosphorylation.

Another group of compounds that have what appear to be "anti-thyroxine" effects are the blockers of the actions or effects of the catechol-amines. Because catecholamine calorigenesis is exaggerated in thyrotoxic subjects, there has been considerable discussion of the possibility that the calorigenic effects of thyroxine are really mediated through the catechol-amines. This hypothesis is supported by the experiments of Holtkamp and Heming (1953) that demonstrate that adrenergic blockers like Dibenzyline prevent small doses of thyroxine from raising the BMR, but not large doses. Similar experiments of Brewster *et al.* (1956) have been criticized because enough Dibenzyline was given to partially paralyze the animals, and so might have lowered the BMR by decreasing muscle tone (Surtshin *et al.*, 1957).

Mg^{2+} ions are antagonists to the actions or effects of the thyroid hor-mones, both *in vitro* and *in vivo*. A diet high in Mg prevents the mito-chondrial uncoupling seen in rats after administration of thyroid hormones; a diet low in Mg promotes the uncoupling action of thyroid hormone. The effect of administering Mg^{2+} salts upon thyrotoxicosis is controversial, some finding a decrease in BMR and heart rate, others finding no decrease in BMR or change in negative nitrogen and phosphate balances.

In vitro, Mg^{2+} ions antagonize the actions of thyroxine upon isolated mitochondria (Bain, 1954), and it is necessary to limit Mg^{2+} concentration to demonstrate uncoupling and the secondary hormonal stimulation of protein synthesis that is mediated by mitochondria (see Chap. 8). These antagonisms are specific in that Mg^{2+} ions do not act similarly against other phenolic uncoupling agents like DNP or the salicylates. That specificity is one of the pieces of evidence that the thyroid hormones act on mitochondria by a mechanism different from that of DNP. It is not clear if Mg deficiency changes respiratory control directly or if the deficiency allows normal amounts of thyroid hormones to change respiratory control, or even if the thyroid hor-mones themselves act through Mg^{2+}.

Evidence has been presented tht the Mg^{2+}-thyroid antagonism is a reflection of a binding between Mg^{2+} and thyroxine. Insoluble complexes can form; however, if direct binding exists, the Mg^{2+}-activated enzymes should be inhibited by thyroid hormones. Some are; e.g., the creatine phosphokinase of rabbit skeletal muscle is inhibited by L-T_4 *in vivo* and *in vitro*, the latter on kinetic grounds via the formation of a complex of Mg^{2+} ions in competition with enzyme or substrate. On the other hand, the Mg^{2+}-requiring hexokinase of rat muscle is not inhibited by L-T_4 *in vitro* and is actually activated by L-T_4 administration *in vivo*.

The thyroid state of a subject influences the distribution of Mg in the body. Myxedematous patients excrete large amounts of Mg^{2+} in their urine promptly after hormone administration. Circulating Mg^{2+} concentrations are also affected. The concentration of Mg^{2+} in plasma is low in hyperthyroidism and high in hypothyroidism, and Mg balance is positive in hyperthyroidism and negative in hypothyroidism. There is still controversy about the state of free Mg^{2+} in the plasma. Some investigators have reported that the concentration of unbound, diffusible Mg^{2+}—which is probably the effective moiety of total Mg^{2+} in the plasma—is low in hyperthyroidism and high in hypothyroidism, sometimes in the absence of large changes in total Mg^{2+}; however, others find no such changes. Cellular exchangeable Mg^{2+} is very low in hypothyroidism but normal in hyperthyroidism (Jones *et al.*, 1966).

BARBITURATES

The barbiturates have both accelerating and inhibiting effects on energy transformations, the resultant is usually on the inhibitory side. Furthermore, some of these compounds reach the central nervous system selectively, and so do not have prominent actions on the visceral tissues.

Barbiturates, as exemplified by Amytal (Amobarbital, 5-ethyl-5-iso-amyl barbituric acid, Fig. 14–2), are inhibitors of electron-transport in the mitochondrion. In general, they inhibit at a flavoprotein site in the region of the NADH-reductase, and so inhibit respiration when NAD-dependent substrates are oxidized, but not succinate (see Chap. 3). That inhibitory action would be expected to decrease respiration and energy release *in vivo*, but the degree to which it would lower the BMR depends upon how much of the respiring tissue mass the barbiturate reaches. Brain respiration alone accounts for more than 2.6 per cent, perhaps up to 10 per cent, of the total body uptake of oxygen (see p. 6).

However, the barbiturates have another effect on energy release that is more general; they are hypnotics and anesthetics that decrease muscle tone through the depression of the function of the central nervous system. Just as in sleep, a general relaxant effect makes the measured metabolic rate more "basal," usually about 15 to 20 per cent below the level of the BMR that is measured in an awake subject.

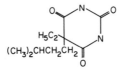

AMYTAL

Figure 14–2. The structure of amytal (amobarbital; 5-ethyl-5-isoamyl barbituric acid).

The barbiturates are also uncoupling agents, and act on mitochondria very much like DNP, although perhaps through a different mechanism (Brody and Bain, 1951, 1954; Brody, 1955; and Aldridge and Parker, 1960). Mg^{2+} ions do not reverse the uncoupling action of barbiturates or DNP. Barbiturates stimulate State 4 respiration and depress phosphorylative efficiency *in vitro,* and act on all of the intermediates of the Krebs cycle. Liver and brain mitochondria are sensitive to barbiturates, those from brain more than liver.

There are some problems in interpreting the effects of the barbiturates *in vivo* on the basis of their uncoupling actions *in vitro.* For one thing, why aren't all uncoupling agents hypnotics or anesthetics? That seems to be answered by the lack of penetration of the usual uncoupling agents into the central nervous system. For another, why aren't the barbiturates calorigenic like other uncoupling agents? Presumably because the barbiturates have the added property of being specific inhibitors of mitochondrial respiration. Why don't the barbiturates deplete high-energy $\sim P$ compounds and increase P_i in the central nervous system if they act there by uncoupling? They actually increase $\sim P$ and lower P_i (Buchel and McIlwain, 1950). Perhaps this is because the utilization of $\sim P$ is depressed even more than its generation (Brody, 1955). The failure of barbiturates administered *in vivo* to depress the P:O ratios of mitochondria subsequently prepared from rat brains is answered by invoking a "washing-out" of the barbiturates during the isolation of the mitochondria. Lastly, there seems to be no correlation between the uncoupling potency of barbiturates *in vitro* and their hypnotic potency *in vivo*, although that correlation is better in the group of barbiturates that act as convulsants, not depressants. Again, these objections need interpretation in the light of data on tissue contents after administration *in vivo.*

There are, however, some more indirect pieces of evidence in which mitochondrial uncoupling is involved in the *in vivo* effects of the barbiturates. The hypnotic effects depend upon the thyroid state of the subject, very much like the calorigenic effects of uncoupling agents do. Several known uncoupling agents potentiate barbiturate effects *in vivo.* Here we run into yet another complicating phenomenon: the effect of Drug I can be potentiated or inhibited by the previous administration of Drug II that induces changes in the amount of enzymes that metabolize Drug I (see Conney, 1969). The barbiturates are by now well-known stimulators of enzyme induction, and can cause striking changes in the metabolism of either administered compounds or normal body constituents. Chronic administration of a barbiturate (phenobarbital) increases the binding of thyroxine in liver cells, and the hepatic turnover of thyroxine through deiodination and biliary excretion, which may affect hormone action (Oppenheimer *et al.*, 1968). Conversely the thyroid hormones are also promoters of (respiratory) enzyme induction, and depending upon the species, they modify the effects of the barbiturates by altering the sensitivity of the tissue responses to barbiturates (as the hormone controls the effects of DNP), or by altering the amounts of barbiturates that reach or stay in the target tissues.

A number of uncoupling agents (DNP, aureomycin, 2,4-dichlorophenoxyacetic acid) increase the anesthetic potency of amobarbital (subanesthetic doses become anesthetic) and prolong the sleeping time after secobarbital up to 300 per cent in mice and frogs. The potentiating effect of DNP in these studies is not mediated by changes in the rate of detoxification of the barbiturate, as measured by brain and blood levels (Brody and Killam, 1952); however, in rats, multiple injections of L-thyroxine increase the duration of action of hexobarbital, and decrease the activity of microsomal enzymes in the liver that remove hexobarbital (Conney and Garren, 1961). In mice, the feeding of desiccated thyroid potentiates the effects of pentobarbital and thiopental, and delays drug removal from the brain; hypothyroidism due to the feeding of propylthiouracil decreases the responses to pentobarbital (but not to thiopental) and accelerates its removal from tissues. It is postulated that the hormone induces increased activity of enzymes that destroy pentobarbital, but may also decrease the sensitivity of the nervous system (Prange and Lipton, 1966); however, in rats the thyroid state has no effect on sensitivity, but either hyper- or hypothyroidism delays the removal of the drug and prolongs its action. A possible involvement of thyroid hormones with the ability of rats to withstand cold environments and the decrease of body temperature after injection of pentobarbital is pointed out by Gemmill and Browning (1962).

In man, it is not clear that the thyroid state affects the response to barbiturates, or that barbiturates affect the BMR. Indeed, there are suggestions that barbiturates are of value in preparing patients for measurements of the BMR—to make them more "basal" (Satoskar et al., 1965). Whether or not synergisms exist in man similar to those seen between other uncoupling agents and thyroid hormones in experimental animals remains to be seen.

ENVIRONMENTAL TEMPERATURE

Hypometabolism is seen after exposure of animals or man to either warm environments or very cold ones. Some mammals hibernate when exposed to moderately cold environments; their metabolic rates are markedly depressed. Because these phenomena also involve decreases in body temperature, they are discussed in the next chapter.

General References: Hypometabolism due to extreme cold is discussed in a symposium edited by Musacchia and Saunders (1969).

15

TEMPERATURE REGULATION

The body temperatures of mammals are maintained at a constant level regardless of the temperature of the environment, within a relatively narrow range of environmental temperatures: mammals are homeotherms. Temperature is regulated through a series of compensating mechanisms that balance heat production against heat loss. Those mechanisms seem to be initiated by a hypothalamic area that is sensitive to the temperature of the blood. Thus exposure of a mammal to a cold environment rapidly produces neural and hormonal responses that initiate peripheral vasoconstriction that decreases the loss of heat, and shivering that increases heat production. Exposure to warm environments causes sweating that accelerates heat loss and later causes a depression of heat production. Two sets of hormones are involved chiefly: the catecholamines that effect rapid transient changes, and the thyroid hormones that effect slower and more lasting changes, but perhaps also act rapidly. Other hormones are probably involved as well, e.g., the steroids of the adrenal cortex.

From the point of view of energy transformations we are most concerned with the mechanisms whereby heat production is increased or decreased: mitochondrial metabolism is the main source of heat (an additional discussion, concerned mainly with the role of lipid metabolism, will be found in "Lipid Metabolism" by Masoro).

How the mitochondrion produces more or less heat seems relatively simple when an artificial *in vitro* system is considered. In that thermodynamic system there are only two alternatives for the fate of the free energy liberated by oxidations. Either energy-rich bonds are generated and *stored* in the form of $\sim$P-bonds (in ATP or a phosphate-acceptor, usually glucose-6-phosphate) or in ion gradients inside the mitochondrion, or the energy is not transformed because of uncoupling and heat is generated. In the coupled system, the free energy of oxidation is distributed about 3:2 in favor of ∼P-bonds versus heat. In the uncoupled system all the free energy is transformed to heat.

The intact animal receiving an uncoupling agent has been considered to be analogous to the uncoupled mitochondrial system *in vitro*. The intact animal obviously produces excess heat after being treated with DNP, and the body temperature may rise high enough to be lethal. The rise in heat production *in vivo* is considered to be due to uncoupling and loss of efficiency of mitochondrial energy transfer. Conversely, in the long-term acclimation of animals to cold temperatures, which involves increased heat production, it has seemed reasonable to invoke an uncoupling mechanism, although the evidence for its actual occurrence has been disputed.

However, the simple and direct interpretation of these *in vivo* phenomena on the basis of *in vitro* experiments on isolated mitochondria can be questioned on several grounds. Teleologically it is not reasonable that an animal should sacrifice its metabolic efficiency to produce heat in acclimatizing itself to cold (Hoch, 1962a). Exposure to a cold environment raises the demands not only for body heat, but also for a more rapid turnover of body components, with a concomitant acceleration of synthetic processes that require utilizable energy. The process of loose-coupling, whereby mitochondrial oxidations are stimulated with only a slight loss of efficiency, has seemed more compatible with the animal's successful acclimation and with most of the data on such animals' tissues. Loose-coupling can be induced through the presence of agents capable of uncoupling at high concentrations, and represents a rapid, short-term response to cold through a rise in the specific activity of the oxidative enzymes. For more lasting adaptation, an increase in the amount of oxidative enzymes also accounts for greater heat production even with the efficiency of energy-transfer remaining normal. There is good evidence that this occurs, too.

On thermodynamic grounds it has been pointed out (Beyer, 1963) that in the normal animal heat is produced not only directly from the 60 per cent of the energy from mitochondrial oxidations liberated directly as heat, but also eventually from the other 40 per cent originally transformed into ∼P-bonds or nonphosphorylated energy-rich bonds. In an animal in the steady state (at rest, and not gaining or loosing weight), the energy-rich bonds produced are turned over continuously in the course of utilization by the cell. If there were no such turnover, ATP would accumulate progressively, or its energy-rich bonds would accumulate in the sense of their increasing the length of the energy-storing polymers of the cell (fats, glycogen, proteins) through synthetic processes, or ions would continue to accumulate in mitochondria. But the animal is not gaining weight, and in the steady state no such storage takes place. *Ergo*, even the 40 per cent of liberated energy originally transformed to utilizable forms is eventually used and transformed to heat.

Since the pathway makes no difference in a thermodynamic system, the normal live animal doing no external work operates as if its mitochondria produced only heat. The isolated mitochondrion and the intact animal are different thermodynamic systems, and it may not be permissible to extrapolate from the simpler to the more complex system without taking that into regard. (Inherent in all these considerations is an assumption that the entropy of the two systems remains the same, and that is not only far from proved but difficult to prove.)

How, then, does the living system increase its production of heat under the influence of uncoupling agents if it operates normally as if it were already uncoupled? *Through acceleration of oxidation.* Uncoupling agents accelerate oxidation by depressing respiratory control. In the phenomenon of loose-coupling, accelerated oxidation is accompanied by a normal efficiency of energy conversion. Physiologic agents (like thyroid hormones) can accelerate oxidation through loose-coupling, but also indirectly through their stimulating effects on the synthesis of the proteins of the oxidative apparatus. Conversely heat production can be diminished by a slowing of oxidation. The exogenous agents that inhibit electron-transport can act in this manner. Endogeneous agencies (for instance, a lack of thyroid hormones) decelerate oxidation through increases in mitochondrial respiratory control, and also indirectly through a depression of the synthesis of respiratory enzymes.

The importance of oxidative rate to the production of heat as compared with the efficiency of energy-transfer has recently been stressed on somewhat different grounds. Even assuming that the efficiency of oxidations in the intact animal can be considered in the same terms as in the mitochondrion *in vitro*, it has been calculated that only about 25 per cent of the heat generated from acetate, for instance, is transduced to useful energy (Prusiner *et al.*, 1968a). If such a system is uncoupled, only that 25 per cent will contribute to additional heat production. The difference between the 25 per cent calculated in these studies and the 40 per cent or more referred to on page 44 as the efficiency of energy transduction arises from Prusiner *et al.* using enthalpy values rather than free energy values to calculate the heat produced by the hydrolysis of ATP to ADP and P_i. The enthalpy change is -4.7 kcal per mole, about 30 per cent lower than the free energy value. These calculations are supported by direct calorimetric studies with non-phosphorylating submitochondrial particles (Poe *et al.*, 1967). Prusiner *et al.* conclude that heat is generated mainly through respiration and not the hydrolysis of ATP, and that mechanisms for the control of heat production are best described in relation to the control of respiration, which is regulated by the presence of substrate and phosphate acceptors.

The special role of the brown adipose tissue, with its high content of mitochondria, in evolving heat in mammals adapted to prolonged cold-exposure is discussed on page 69.

INCREASED HEAT PRODUCTION

Adaptation to Cold

The experimental conditions of exposure to cold environments have been very varied. "Acclimation" has been used to designate the changes evoked by long-term exposure to environments at constant (low) temperatures (Depocas, 1961). Laboratory or wild animals and man have been studied during and after exposure to winter conditions in places like Canada and Scandinavia where extreme cold might occasionally prevail; the stress in man has been kept relatively uniform, by limiting his garments and bedclothes, as well as regulating his physical activities. A special set of conditions has become of interest in certain surgical procedures where subjects are kept at low body temperatures for brief periods by immersion and irrigation. It is difficult to find common parameters for all these conditions, and it is not surprising that different animals (coming as they do from different natural environments) should respond to cold variously.

Small rodents have usually been studied because of their availability and general use in metabolic experiments. Rats and mice can survive at temperatures below 30° and down to about 5 to 10°C by rapidly increasing their heat production and then maintaining a high BMR. However, similar animals kept under natural winter conditions do not show a high resting metabolic rate, nor do naturally wild animals, nor does man when kept on a program of increased muscular activity during cold exposure (Wilson, 1966). Wild animals respond to cold differently from laboratory animals, having increased pelage insulation and lower peripheral temperatures. It thus seems clear that there are different types of acclimation.

At temperatures lower than about 5°C, heat loss in laboratory rodents exceeds heat production and the animals become hypothermic and die. In experiments on induced hypothermia under anesthesia, for surgical procedures, the cause of death when the temperature is too low is usually a cardiac arrhythmia (ventricular fibrillation). In the rat, hypothermia decreases the cardiac content of ATP and phosphocreatine, and changes in mitochondrial morphology lead to defects in conduction and alterations in the electrocardiogram (Zimny and Taylor, 1965).

Acclimation to moderately cold temperatures takes between 2 and 4 weeks in rats and mice. A series of compensatory changes in energy transformations occurs which successively employs different mechanisms to increase heat production. Without such changes, the metabolic rate falls exponentially with body temperature in accord with Van't Hoff's law, decreasing about 50 per cent at a body temperature of 28°, and 87 per cent at 18°. An immediate set of responses gives way gradually to a set that persists when cold exposure is prolonged. There is still considerable disagreement about the mechanisms, but the phenomena themselves seem clear.

Shivering thermogenesis occurs immediately upon exposure to cold and persists in rats for a few days or in monkeys for a few weeks when exposure to cold is prolonged. Shivering arises probably directly through nervous stimulation of the central nervous system and perhaps via decreased blood temperature. Shivering acts via an ATPase and produces ADP by the act of muscle contraction. The ADP excess, just as in exercise, accelerates mitochondrial oxidations by shifting respiration to State 3. The increased rate of oxidation produces more heat. Shivering diminishes gradually with continued exposure to cold, but the heat production of the animal continues high; at this state, *metabolic thermogenesis* is said to have taken over. Actually metabolic changes occur very soon after the animal is chilled, and their contribution to heat becomes greater and finally predominates.

Among the earliest changes after cold exposure are increases in endocrine secretions and responses. The catecholamines and hormones of the thyroid-pituitary system are involved. The catecholamines are secreted very promptly, and their calorigenic actions—especially that of norepinephrine—raise the production of heat.

There is a very rapid release of TSH, which reaches a peak in 30 minutes, persists for a few hours, and then declines but still remains above normal levels (Itoh *et al.*, 1966). TSH release is most striking at moderately cold temperatures, as is the resulting release of thyroid hormones from the thyroid gland. At 16°, no T_4 is released; at 6.5 to 11° there is a marked stimulation of T_4 release, but at 2° T_4 release is depressed even below normal levels. At very low temperatures, the activity of the anterior pituitary and the thyroid gland seems to be markedly suppressed (Brown-Grant, 1956), either by the decreased temperature of the glands themselves or by the slowing of the circulation owing to the extreme vasoconstriction effected by the catecholamines.

In the acute phase of cold adaptation, the secretion and release of thyroid hormones occur within the first few hours (Tamada *et al.*, 1965; Slebodzinski, 1962). Exposure to moderate cold (15 to 21°) is more stimulating to thyroid activity than extreme cold (1°) (Brown-Grant *et al.*, 1954). Later, there is continued hyperproduction of thyroxine, but there is also more rapid destruction and excretion. In the early hours, the mitochondria obtained from the livers of cold-exposed rats take up less LT_4 from the media in which they are suspended, indicating that the mitochondria are partly presaturated with thyroid hormone *in vivo* (Tonoue and Matsumoto, 1961). The actions of the large amounts of thyroxine that are released are not yet conclusively demonstrated, but from the results of experiments on thyroid hormone actions certain extrapolations can be ventured. The rapid flooding of the bodies of cold-exposed animals with thyroxine might occasion (1) mitochondrial uncoupling or loose-coupling, (2) potentiation of the action of uncoupling agents, and (3) subsequent synthesis of respiratory enzymes. All three phenomena are reported.

(1) There is evidence for uncoupling in the tissues of cold-exposed

animals. Cold-exposed rats have decreased liver contents of high-energy phosphate compounds and high P_i (Beaton, 1963). The first report of Martius and Hess (1951) on uncoupling by injected thyroid hormone also included a group of euthyroid rats exposed to low temperatures, whose liver mitochondria were similarly uncoupled. Other laboratories (R. E. Smith, 1960a, b) have reported small decreases in P:O ratio that are reversed by washing the mitochondria or by extracting with albumin. Skulachev (1963; Skulachev *et al.*, 1963) observed uncoupling in the skeletal muscle mitochondria of mice exposed to extreme cold for only 2 hours, and the reversal of the uncoupling and swelling upon addition of albumin or ATP. Such findings indicate that some removable substance causes uncoupling in the mitochondria, but do not show the nature of that substance. That it might be thyroxine is suggested by the fact that albumin reverses the action of injected hormone and removes most of the hormone from mitochondria (Hoch and Motta, 1968); it could also be fatty acids. Thyroidectomy appears to repair the uncoupling caused by exposing rats to cold (Lianides and Beyer, 1960a, b).

There are no published direct measurements as yet of thyroxine in the mitochondria of cold-exposed animals. We have recently found that the skeletal muscle mitochondria obtained from rats kept at 15°C for 2 hours contain about four times as much iodine as those from rats kept at 23°C or 5°C; liver mitochondria from cold-exposed rats contain less than the normal amount of iodine. Cold exposure seems to redistribute the hormone, and the accumulation of hormone in muscle mitochondria is in contrast to the very small amount of extra iodine that is found in such mitochondria after injecting the hormone. Perhaps these findings indicate why animals that are exposed to cold do not appear to be simply thyrotoxic; they show increased hormone content in the organ where shivering produces heat via the ATPase mechanism. These very rapid changes seem to be in contrast to other findings at lower temperatures for longer times. All the tissues of rats exposed from 0 to 4°C for 15 days contain more T_3 that those of control rats, increases of +75 per cent in liver, +83 per cent in kidney, and +221 per cent in epididymal fat pad being reported; the T_4 content of these three tissues is slightly elevated, and of the other tissues, unchanged or decreased (Albright *et al.*, 1965). The increased T_3-content of brain (29 per cent), spleen (+80 per cent), and testis (+54 per cent) in these studies is difficult to reconcile with the normal respiration of these three tissues in cold-exposed or thyroid-treated animals.

On the other hand, many laboratories have found no changes in mitochondrial respiratory control or P:O ratio after exposing animals to the cold (Aldridge and Stoner, 1960, 1963; Chaffee *et al.*, 1961; Patkin and Masoro, 1960).

(2) The calorigenic action of the catecholamines is potentiated very rapidly in cold-exposed animals, and continues high with continued exposure. In as early as 40 minutes after cold exposure, injected norepinephrine

and (less so) epinephrine raise heat production more markedly than in control animals (Carlson, 1960; Swanson, 1956, 1957). That a major role of thyroxine in cold adaptation is to increase the calorigenic action of catecholamines was suggested as early as 1942 by Ring. The continued excessive responses with prolonged exposure, together with the persistent increase in the turnover of thyroid hormones, are consistent with this view and have been documented by a number of workers. In addition, it is known that adequately hypothyroid animals respond little if at all to catecholamines, and that they survive for only very short periods of time in the cold, probably by shivering. Some studies report that hypothyroid animals survive for long periods. Probably the disparity arises from different degrees of hypothyroidism. The usual laboratory chows contain thyroxine and iodide in quantities sufficient to increase the metabolism of hypothyroid animals (Leblond and Eartley, 1952). Adequate hypothyroidism is only attained on rations that contain no thyroid-active substances. Part of the difficulty is that the more hypothyroid an animal is, the greater its sensitivity to thyroxine. It takes very little thyroxine to maintain a hypothyroid animal in the cold.

The observations of Sellers and You (1950), that 12.5 $m\mu g$ of $L\text{-}T_4$ per g per day did not restore the low BMR of thyroidectomized rats fed propylthiouracil, yet did restore the slow rise in BMR normally seen during gradual exposure to a cold environment, seem interpretable in the light of the more recent findings. Firstly, the administration of propylthiouracil to thyroidectomized rats antagonizes the peripheral actions or effects of administered thyroxine (see p. 161), thus the dose of $L\text{-}T_4$ given by Sellers and You was in effect even smaller than it appeared. Secondly, the amount of thyroid hormone necessary to restore the depressed sensitivity of hypothyroid rats toward calorigenic agents is much less than that necessary to restore the depressed metabolic rate. Indeed, a dose of $L\text{-}T_4$ (5 $m\mu g$) that does not raise the BMR at all (and perhaps decreases it slightly) restores full sensitivity toward injected DNP within 6 to 18 hours (Hoch, 1965b); intravenous injection of $L\text{-}T_3$ restores the full calorigenic action of the catecholamines within 3 hours (Bray and Godman, 1965), well before any demonstrable rise in BMR due to the thyroid hormone.

Sellers and You concluded that the slow cold-induced increase in metabolic rate (measured at 30°) does not depend on a hyperthyroid state but does depend on the presence of thyroid hormone, and that this increase in metabolic rate is associated with the ability of the animal to survive. It now seems that the sensitivity of the animal to calorigenic agents may be a crucial factor, and that the sensitivity is under the control of the amount of thyroid hormone present in the tissues. The direct action, but not the secondary effects of thyroid hormones, appear to be necessary for survival in the cold under these experimental conditions.

If thyroid hormones are necessary for cold adaptation, does administering thyroid hormones to normal animals or subjects increase adaptation or survival? The question is a practical one for meeting the stresses of cold environments. The answer is not yet clear. In mice, pretreatment with thyroxine slows the rate of fall of body temperature after exposure to a 5° environment; pretreatment with DNP accelerates the temperature drop (Turner, 1946), perhaps because DNP does not potentiate catecholamines. The survival of rats in the cold was not enhanced by T_4 (1 μg per g per day for 7 days) (Sellers *et al.*, 1951), but treatment of rats in another study (Bauman and Turner, 1967) with 30 mμg per g per day for 7 days, or with cortisone, or more effectively with both, increased survival at 4.5°C.

(3) Adaptation of animals to prolonged cold alters the enzyme content and activity of their tissues. Thyroid hormones characteristically increase the amount of cytochromes in mitochondria. Cytochrome *c* is found only in mitochondria, and cytochrome *c* content increases in the livers of cold-exposed rats (Klain, 1963).

Coenzyme Q is a mitochondrial electron-transport component, but is also found in other cell fractions. Cold exposure for 2 weeks raises the coenzyme Q content of rat liver, heart, skeletal muscle, and kidney, but not brain (Beyer *et al.*, 1962). In the liver, mitochondrial coenzyme Q is tripled after 40 days of cold exposure, and the microsomal content of coenzyme Q is increased seven-fold; the accumulation arises from a decreased destruction over the entire period, and an increased synthesis during the second 10 days (Aithal *et al.*, 1968). Administering thyroxine or DNP or cortisone to normal animals also increases tissue and mitochondrial coenzyme Q (Ramasarma *et al.*, 1967; Beyer *et al.*, 1962; Aiyar and Sreenivasan, 1962; Pedersen *et al.*, 1963). Beyer has postulated that coenzyme Q functions as an energy-wasting, heat-producing shunt for electron-transport by bypassing a site of energy conservation; the argument over the role of coenzyme Q in mitochondria is discussed on page 24. An analogous postulate of shunting of electron-transport via the mitochondrial pyridine nucleotide transhydrogenase (Potter, 1958) is weakened by a report of normal transhydrogenase activity in cold-acclimated animals (Hannon and Rosenthal, 1963).

The activities of a number of oxidative enzymes also increase (Hannon, 1960; Depocas, 1961); some of them, like cytochrome oxidase and the substrate-specific dehydrogenases, are mitochondrial enzymes, but others are not. The increase in enzyme activity per gram of tissue, taken together with the observed increase in the mass of visceral organs, like the liver, kidneys, and heart (skeletal muscle mass usually decreases), leads to an extensive repatterning of energy metabolism in cold-acclimated rats and mice. Brown fat depots, with their high content of mitochondria, increase at the expense of white fat, and their respiration accelerates (Ikemoto *et al.*, 1967). Some of the enzymes of glycolysis decrease. There is as yet no clear pattern of altered metabolic pathways that identifies the source of thermogenesis absolutely. Furthermore, it is not clear that all the enzyme changes that occur in rodents are necessary in monkeys that are acclimated to cold. In the monkeys, the only increased enzyme activities that were similar to those seen in rodents were the α-glycerophosphate dehydrogenase activities of brown fat and of liver mitochondria and the succinoxidase of kidney mitochondria; the other enzyme activities did not change at all (Chaffee *et al.*, 1966).

Another early compensatory mechanism is an increase in *food intake* that serves the purpose of supplying an energy source for the accelerated oxidative thermogenesis. Early in cold exposure, the mobilization of oxidable substrates usually results in a loss of body weight. In 1 to 2 weeks food intake increases enough in rodents to satisfy the increased demands on available energy. Monkeys increase their eating until they consume 4 per cent of their body weight per day, without gaining weight; they consume 300 kcal per day, which is equivalent to a man taking in 28,000 kcal per day (Chaffee *et al.*, 1966).

Not all workers in the field of cold acclimation accept a role for the thyroid hormones quite as causal as that outlined here. For instance, Heroux (see discussion of Chaffee *et al.*, 1966) objects to the lack of demonstration that thyroid hormones are responsible for the increased thermogenic capacity or are involved in the mechanisms of nonshivering thermogenesis, which he thinks are in fact "superimposed on the higher resting heat production and not necessarily controlled or affected by thyroxine." Most significantly, the metabolic rates and indirect measurements of thyroid activity are normal when wild or laboratory rats are acclimated to the Canadian winter outdoors in roomier cages. He concludes that "it is wrong to assume that thyroid hormones are directly involved in the metabolic mechanism during cold adaptation" (by "directly involved" he may mean that their *presence* is required). Des Marais and LeMay (1964) find that diacetyl-2,6-diiodohydroquinone, which has antithyroid properties (Serif and Seymour, 1961), has no depressing effect on the raised BMR in cold-acclimated mice; such antithyroid compounds apparently act peripherally to block the calorigenic effects of only administered thyroxine, but not endogenous thyroxine (see p. 160).

In summary, it appears to the author that at present the balance of the evidence is for a causal role for the thyroid hormones in the mechanisms by which certain laboratory rodents adapt to prolonged exposure to cold environments. Further data are needed to explicate that causal role. Cold-exposed animals are not simply hyperthyroid, thus an extrapolation of all the effects of excess thyroid hormone in normal animals to explain the changes in cold-exposed animals is not warranted. Severe thyrotoxicosis would not favor survival at low temperatures, although mildly thyrotoxic animals do survive better at low than at normal temperatures. Perhaps the redistribution of the hormone to muscle mitochondria under conditions of moderate cold indicates a basis for the differences between thyrotoxicosis and acclimation. Certainly the effects of other hormones, such as the steroids of the adrenal cortex, deserve consideration; cortisone modifies the effects of thyroxine (Hoffman *et al.*, 1948; Bastenie and Ermans, 1957).

Fever

Elevated body temperature may be caused either by the inability to lose heat or by the overproduction of heat.

The setting of the hypothalamic thermostat is affected by drugs, hormones, and other substances. Substances that produce a high temperature setting are "*pyrogens*," and they include many proteins and peptides, lipopolysaccharide and lipopolypeptide toxins secreted by bacteria, and perhaps a steroid, etiocholanolone. Etiocholanolone may release an endogenous pyrogen from leukocytes (Bodel and Dillard, 1968). When the hypothalamic thermostat is reset to a higher temperature through the actions of a pyrogen, a complex series of events produces the conservation of body heat and an increase of production of heat. The regulation of the new level of body temperature in man is so precise that administration of a pyrogen achieves about the same body temperature when the subject is in a warm, neutral, or cold external environment.

The increased body temperature after pyrogen administration is accompanied by a rise in the metabolic rate. How much of the increased metabolism is due to accelerated enzymatic rates produced by increased temperature, and how much is due to alteration in mitochondrial respiratory control, is not clear. Bacterial pyrogens do not directly uncouple oxidative phosphorylation or accelerate State 4 respiration *in vitro* or *in vivo*, but they do induce a change in mitochondrial energy transformations that involves an increased turnover of terminal phosphate residues in rat-liver ATP, and that change is distinct from the action of DNP (Ogata, 1962). Mitochondrial respiratory control does appear to be involved somehow in the mechanisms of fever production. For instance, hypothyroid patients occasionally fail to respond to bacterial infections with a normal febrile rise in temperature. As detailed in Chapter 8, the thyroid hormones control the respiratory responses of mitochondria.

Administration of uncoupling agents to humans or to other mammals produces fever when the dose is large enough and the external temperature is not too low. Fever is one of the prominent signs of poisoning with such agents; however, fevers are also caused by drugs that are not uncoupling agents (Borison and Clark, 1967), and all uncoupling agents do not cause fevers. (Some of the barbiturates uncouple *in vitro*, but their inhibiting action on the nervous system and on tissue respirations seems to predominate *in vivo;* see page 165.)

Among the uncoupling agents that produce fever, the salicylates occupy a curious position. Salicylate poisoning, as noted in Chapter 12, is accompanied by fever and a high metabolic rate, which seem to be a reflection of peripheral hyperoxidation and uncoupling. But lower doses of the salicylates have an antifebrile effect. The mechanism of the antifebrile effect is not yet clear despite much study (see M. J. H. Smith, 1963) and may involve either a peripheral action or a central action upon the hypothalamus.

The uncoupling agents like DNP or pentachlorophenol have no antipyretic actions, but phenol itself lowers body temperature below normal levels.

DEPRESSED HEAT-PRODUCTION

Adaptation to Heat

Mammals exposed to heat for long periods gradually adapt by evolving less heat in their tissues. The adaptation seems to involve the thyroid hormones. Hot environments suppress the synthesis of thyroid hormones by the thyroid gland (Fregly, 1967). Heat acts rapidly on the gland, 1 to 3 days in a warm environment halving the release of the hormone in sheep (Slebodzinski, 1962). Whether comparably rapid increases in mitochondrial respiratory control are induced by the rapid early withdrawal of thyroid hormones is not known. The long-range adaptation to heat suggests that the relative lack of hormone depresses the rate of protein synthesis and results in the flavin- and cytochrome-deficient mitochondria characteristically seen in hypothyroidism. Such mitochondria evolve heat less than normally. Hypothyroidism favors the survival of animals in warm environments, and hypothyroid humans are typically comfortable at temperatures that are too warm for euthyroid subjects.

Another mechanism for depressing heat production also depends upon the relative hypothyroidism of heat-exposed animals, but it comes into play earlier than the diminished synthesis of respiratory enzymes. The calorigenic action of the catecholamines is decreased. In hypothyroid animals, the rapid calorigenic responses to catecholamines are depressed or absent. Thus stimuli that normally evoke catecholamine secretion and subsequent heat production do not produce as much heat in heat-acclimated animals.

Humans exposed to warmer climates do not necessarily show lower basal metabolic rates, nor do populations that live in the tropics. Just as with cold-acclimation, heat-acclimation is probably effected by several mechanisms, and it is difficult to assign the process in humans.

Hibernation

Hibernating mammals exhibit a marked decrease in metabolic rate and body temperature. The mechanisms that induce hibernation are not yet understood (see Lyman and Chatfield, 1955; South and House, 1967), and there is disagreement over the roles of humoral or neurogenic processes. Since hibernation is species-specific, it would seem reasonable to expect that the tissue apparatus for energy transformation is different in hibernators than in animals that do not hibernate. This seems to be the case; if it were not, the possibility might exist that any mammal could be induced to hibernate upon administration of the proper "hibernating hormone."

In hibernation, the metabolic rate is about 1 to 3.3 per cent of that in the same animals when awake. Such a degree of depression might arise either from a decreased and controlled specific activity of respiration in the tissues, or from a decreased number of respiratory enzymes. The observation that hibernators can be roused within minutes or hours almost certainly speaks against any marked change in the protein composition of the mitochondria or other respiratory systems—there just is not enough time to synthesize new respiratory enzymes to account for the great burst of oxidative activity that heralds arousal. So the decreased oxidation in hibernation must be due to a suppression of enzyme activity. The capacity for oxidation is intact and even increased (Chaffee *et al.*, 1961; Anthony *et al.*, 1967).

The changes in the tissues of hibernators are superimposed on changes owing to exposure to cold—since hibernation is induced by putting the animals into a cold environment. In the golden hamster, the liver mitochondria obtained from either active or hibernating cold-exposed animals show increased activities of a number of enzymes of the electron-transport system but normal P:O ratios (Chaffee *et al.*, 1961). In contrast, no major mitochondrial changes occur in rats exposed under the same circumstances. The mitochondria of hibernators contain a temperature-sensitive succinoxidase system that has a lower activity when measured at 7° than that observed in active cold-exposed animals or control animals. The appearance of a respiratory system with a higher Q_{10} may represent a basis for the sharp decrease in oxidative rates in hibernation. It is also of interest that mitochondria from cold-acclimatized and hibernating hamsters differ from the controls in being of a deeper red-brown color, probably owing to a greater content of cytochromes.

The process of arousal from hibernation features a marked and rapid production of heat that raises body temperature from a level close to the ambient low temperature to normal. The increased enzyme capacity of the mitochondria is used for the oxidative burst, and the contribution of the mitochondria of the brown body fat is particularly prominent; whether this oxidation is efficient or not is still controversial (Correspondent, 1968). The ATP of the brown fat in bats aroused from hibernation decreases rapidly (Dryer and Paulsrud, 1966), reflecting either increased use or decreased synthesis. The temperature-sensitive suppression of succinoxidase activity in hibernating hamster liver mitochondria apparently disappears very rapidly upon awakening, perhaps accounting for the release of respiratory inhibition (Chaffee, 1962).

General References: The role of lipid metabolism in thermoregulation is discussed by Masoro (1966, 1968). There is extensive and frequent coverage of the phenomena and mechanisms of adaptation to cold (R. E. Smith, 1960c, 1964, 1966; Hardy, 1961; R. E. Smith and Fairhurst, 1958; R. E. Smith and Hoijer, 1962; and Hart, 1969).

16

FULMINANT
HYPERTHERMIA
AS A BASIC
PROBLEM

There are still many unsolved problems in the field of energy transformations, not only on a fundamental level as regards mechanisms, but on a practical, applied, and phenomenological level. Indeed, here as elsewhere the solution to the obvious practical questions will probably be arrived at through studies that have no obvious relationship and are aimed at much more basic problems.

In the last decade, a catastrophic syndrome associated with general anesthesia for surgery has been reported and editoralized upon (Martin, 1968; Editorial, 1966; Saidman *et al.*, 1964; Relton *et al.*, 1966; Trey *et al.*, 1968). It has been called fulminant hyperthermia or malignant hyperpyrexia. Its features are an accelerating rate of O_2 consumption and CO_2 production, high fever, and muscular rigidity appearing after general anesthesia has been started; a resistance to treatment; and a fatal outcome in the majority (70 per cent) of cases. The victims seem to be excellent anesthetic risks, and the surgery itself has not been formidable.

Certain features of fulminant hyperthermia have been common to all the cases, others common to most. The most prevalent sign has been a generalized muscle rigidity, which precedes any recognized fever. The rigidity sometimes appears early after the administration of a usual dose of succinylcholine, an agent that is used to produce muscle relaxation. In experiments on a group of pigs under halothane (Fluothane; 2-Br,-2-Cl-1,1,1,-trifluoro-ethane) anesthesia, 100 mg of intravenously administered succinylcholine chloride immediately produces a generalized convulsion, generalized muscle rigidity that resembles *rigor mortis*, excessive CO_2 output, and body temperatures reaching 109° (Hall *et al.*, 1966); the pigs were litter mates. The pigs which were not litter mates, when treated identically, had no untoward responses at all. Succinylcholine is a calorigenic agent (Hemingway *et al.*, 1964) and may act primarily on skeletal muscles, even in doses that produce total paralysis (Muldoon and Theye, 1969).

However, fulminant hyperthermia is also observed in patients who have not received succinylcholine, at a state in the course of anesthesia later than in those patients who have. The suggestion of a genetically determined intrinsic cause for fulminant hyperthermia is reinforced by a report (Denborough *et al.*, 1962) on a local family in which there were 10 deaths attributable to general anesthesia in a pattern compatible with inheritance due to an incompletely penetrant gene(s). The family's condition was brought to notice in the history given by a 21-year-old man undergoing general N_2O-halothane anesthesia for correction of a compound fracture. Within 20 minutes of induction of anesthesia, he was hypotensive, hot, and sweaty; fortunately he survived because of the caution used in anesthetizing him. Subsequently he underwent spinal anesthesia without incident. His family members had succumbed during anesthesias with ethyl chloride and ether. Neither in him nor in other members of the family was any evidence found for an inherited anomaly. Furthermore, the genetic linkage suggested by the studies on pigs is not necessary for producing hyperthermia; pigs of the same breed and from the same farm, but not siblings, show an 18 per cent incidence of hyperthermia during halothane anesthesia without succinylcholine (Harrison *et al.*, 1968), suggesting a dietary or environmental influence.

Almost all the patients showing fulminant hyperthermia were anesthetized with fluorinated hydrocarbon agents (halothane or methoxyflurane); however, administration of halothane to the vast majority of patients and to experimental animals does not produce any hyperthermia but actually depresses the metabolic rate some 20 per cent below basal levels.

The cause of fulminant hyperthermia is not known. Wilson *et al.* (1966, 1967) note the similarity between the syndrome and the features of DNP-poisoning in laboratory animals. In dogs anesthetized with halothane or pentobarbital and then injected with DNP, 5 μg per g, there are significantly greater temperature rises in the halothane series, and about half of those animals die from the hyperthermia. On the other hand, DNP does not produce hyperpyrexia or muscle rigidity in chickens (Viguera and Conn, 1967). In early experiments in Heymans' laboratory (Heymans and Reginers, 1926), the hyperthermia and hypermetabolism caused by intravenous injection of methylene blue were not affected by administering ether, so uncoupling agents may be specifically synergistic with halogenated hydrocarbon anesthetics. The role of accelerated oxidations is stressed, rather than

ATP-hydrolysis, in accounting for the great amounts of heat evolved in anesthetic hyperthermia (Wang *et al.*, 1969). Wilson *et al.* suggest that malignant hyperthermia in humans may be "a disturbance of biologic oxidation representing an individual metabolic defect which has been "triggered" by an anesthetic agent or technique." Other authors stress the need for early recognition of the syndrome so that anesthesia can be stopped, refrigeration applied to remove the excess heat, and other supportive measures started.

The features of fulminant hyperthermia certainly do fit the picture of the uncoupling of oxidative phosphorylation down to the premature appearance of *rigor mortis*. The experiments of Wilson *et al.*, seem to demonstrate a synergism between administered DNP and halothane but leave unexplained the nature of the "individual metabolic defect" that underlies and sensitizes the response to DNP. From our discussion of the sensitizing actions of the thyroid hormones to a large variety of uncoupling agents (Chap. 13), one cannot help but suggest that the metabolic defect might be a mild hyperthyroidism; our recent finding that exposure of rats to a moderately cold environment very rapidly elevates the amount of iodine in their skeletal muscle mitochondria (p. 173) suggests a basis for the muscle rigidity in fulminant hyperthermia. The one piece of direct evidence on patients is so far negative: in a report (Stephen, 1967) on a series of 12 cases, a serum protein-bound iodine estimate in one patient at the time of hyperthermia was within normal limits. Some cases of thyroid crisis after thyroidectomy (Pemberton, 1936) may be an example of fulminant hyperpyrexia from the point of view of the anesthetist. About half the cases of thyroid crisis in a series occurred after surgery, most of which occurred after thyroidectomy. These crises were manifest between 4 and 16 hours after the operation (McArthur *et al.*, 1947). In that series, fatal outcome was correlated with severity of thyrotoxicosis, which does not fit at all with the hyperthermia during anesthesia, in which the patients were never observed to be thyrotoxic. Nor has muscle rigidity been noted as a feature of thyrotoxic crisis. In a fatal case of hyperpyrexia, cytoplasmic glycogen was absent in skeletal muscle, myocardium, and liver cytoplasm (Thut and Davenport, 1966); cytoplasmic glycogen is absent in thyrotoxicosis.

Fulminant hyperthermia, if it is an example of thyroid hormone sensitization, should involve the use of an agent with uncoupling properties during the anesthesia procedure that is the precipitating cause. Obviously no one agent has been identified to date as being the common denominator; however, the very nature of anesthesia may involve the use of agents that have the capacity to uncouple oxidative phosphorylation. Pauling (1961, 1964) suggests that anesthetic gases form "hydrate micro-crystals" with water and the polar side groups of the peptide chains in proteins, thereby altering ion movements or the catalytic centers of enzymes. These properties are related to the electrical polarizability of the anesthetic gas. Other investigators (see Feinstein, 1964) have shown that the potency of gaseous

as well as local anesthetics is strikingly correlated with their ability to penetrate into phospholipid films. The lipid portion of the nerve membrane is
more important for maintaining function than is the protein portion (Tobias,
1960). Quastel (1952) proposes that anesthetics interfere with the reoxidation of NADH, which in turn depresses function of the citric acid cycle;
such an action places the site of the primary event of anesthesia in the
mitochondrion. Whatever the mechanism, it seems that gaseous and local
anesthetics have a common action on the brain (Frank and Sanders, 1963).
Recent studies on the action of local anesthetics on mitochondria show that
ion movements decrease membrane buffer capacity and increase the sensitivity of the membrane to pH gradients (Chance *et al.*, 1968b). As noted in
Chapter 13 (Glaubach and Pick, 1930, 1931), cocaine, procaine, and Novo-
cain are calorigenic agents, and thyroid treatment sensitizes subjects
toward their effects.

In other words, by the chemical theory or by the chemiosmotic theory,
anesthetic gases can act on mitochondrial membranes in much the same
way as an uncoupling agent does.

Indeed, there is direct evidence that some anesthetic agents do act as
uncoupling agents. Halothane added to rat-liver mitochondria uncouples
oxidative phosphorylation at all three sites with various substrates (Snodgrass and Piras, 1966). At 10 mM uncoupling is complete. Increases of
State 4 respiration account for the loss of respiratory control. This action,
which halothane has in common with other hydrocarbons, such as CCl_4,
chloroform, benzene, and diethylether, is attributed to the water solubility
of the agent. After oral administration of halothane, no impairment in
mitochondrial function is detected, nor is the mitochondrial calcium
accumulation that is seen after CCl_4 administration (Thiers *et al.*, 1960)
observed after halothane. Species differences or inadequate oral dosage are
invoked to account for the lack of observable uncoupling with halothane
administered *in vivo;* one might add to those reasons the possibility that
halothane is washed out of the liver mitochondria during their preparation.

In dogs anesthetized with halothane, chloroform, or ether, the myocardial mitochondria show decreased P:O ratios, and the halothane effect
is less than that caused by chloroform, but more than ether (Barkan and
Mistakopulo, 1966). Human liver cells perfused with media containing
higher concentrations of halothane than those used for light anesthesia are
vacuolized (Corssen *et al.*, 1966). Furthermore, the energy-requiring process
of sodium transport in the isolated toad bladder is inhibited proportionately
to the amount of halothane applied (Andersen, 1966).

The associations presented in this chapter may be no more than coin-
cidences, but the evidence from a number of experimental areas seems to
converge upon the conclusion that a sensitized action of an uncoupling
agent is the cause of fulminant hyperthermia. Whether that is so, and what
to do about it if it is so, should be solvable by the appropriate experimentation.

BIBLIOGRAPHY

Abelin, I. (1947). Über den Einflusz des 4-Methylthiouracils auf die Wirkung des Schild-drüsenhormons. Arch. Int. Pharmacodyn. *75:* 187–209.

Abelin, I., and Goldstein, M. (1954). Zur Frage der spezifisch-dynamischen Wirkung des Eiweiszes. Helv. Physiol. Pharmacol. Acta *12:* C59–C60.

Abelin, I., and Goldstein, M. (1955). Über die Mitbeteiligung des Adrenalin und seiner Derivate um der spezifisch-dynamischen Eiweiszwirkung beim Menschen. Biochem. Z. *327:* 72–84.

Aithal, H. N., Joshi, V. C., and Ramasarma, T. (1968). Effect of cold exposure on the metabolism of ubiquinone in the rat. Biochim. Biophys. Acta *162:* 66–72.

Aiyar, A. S., and Sreenivasan, A. (1962). Content and intracellular distribution of ubiqui-none in the rat in experimental thyrotoxicosis. Biochem. J. *82:* 182–184.

Albright, E. C., Heninger, R. W., and Larson, F. C. (1965). Effect of cold-induced hyper-thyroidism on iodine-containing compounds of extrathyroid tissues. *In* Cassano, C., and Andreoli, M. (eds.): Current Topics in Thyroid Research, Proc. Fifth Int. Thyroid Conf., Rome, 1965. New York, Academic Press, Inc., pp. 346–355.

Aldridge, W. N., and Parker, V. H. (1960). Barbiturates and oxidative phosphorylation. Biochem. J. *76:* 47–56.

Aldridge, W. N., and Stoner, H. B. (1960). The behaviour of liver mitochondria isolated from rats with different body temperatures after limb ischaemia or after injection of 3,5-dinitro-*o*-cresol. Biochem. J. *74:* 148–154.

Aldridge, W. N., and Stoner, H. B. (1963). Oxidative phosphorylation in liver mitochondria from cold-acclimated rats. Biochim. Biophys. Acta *68:* 736–739.

Aldridge, W. N., and Street, B. W. (1968). Mitochondria from brown adipose tissue. Biochem. J. *107:* 315–317.

Alwall, N. (1936). Über die Wirkung der Dinitrophenole auf die Tierischen Oxydations-prozesse. Skand. Arch. Physiol. (Suppl.) *72:* 1–117.

Alwall, N. (1939). Steigerung des Citratstoffwechsel *in vitro* durch Salicylsaüre. Acta Med. Scand. *102:* 390–395.

Alwall, N., and Scheff-Pfeifer, I. (1936). Über den Synergismus Dinitrophenol-Thyroxin und Methylenblau-Thyroxin am künstlich durchströmten isolierten Hundebein. Arch. Exp. Path. Pharmakol. *184:* 296–304.

Alwall, N., and Sylvan, S. (1937). Die Stoffwechselwirkung von Thyroxin und Dinitro-phenol in der Narkose. Skand. Arch. Physiol. *77:* 5–6.

Alwall, N., and Sylvan, S. (1939). Über den Einfluss von Urethan auf die Stoffwechsel-wirkung von Thyroxin und Dinitrophenol. Acta Med. Scand. *101:* 290–297.

Ambrose, A. M. (1942). Some toxicological and pharmacological studies on 3,5-dinitro-*o*-cresol. J. Pharmacol. Exp. Ther. *76:* 245–251.

Andersen, N. B. (1966). Effect of general anesthetics on sodium transport in the isolated toad bladder. Anesthesiology *27:* 304–310.

Andik, I., Balogh, L., Donhoffer, S., and Mestyan, G. (1949a). The thyroxine sensitivity of normal and thyroidectomized rats. Experientia *5:* 211–212.

Andik, I., Balogh, L., and Donhoffer, S. (1949b). The effect of thyroxine in thyroidectomized rats treated with methylthiouracil. Experientia *5:* 249–250.

Anthony, A., Munro, D. W., and Stere, A. (1967). Effect of hibernation on respiration and oxidative phosphorylation in chipmunk liver homogenates. Proc. Penna. Acad. Sci. *40:* 43–46.

Arbogast, B., and Hoch, F. L. (1968). Iodine content of submitochondrial particles prepared from rat liver by drastic sonication. FEBS Letters *1:* 315–316.

Askew, B. M. (1962). Hyperpyrexia as a contributory factor in the toxicity of amphetamine to aggregated mice. Brit. J. Pharmacol. *19:* 245–257.

Astwood, E. B., Raben, M. S., Rosenberg, I. N., and Westermeyer, B. W. (1953). Metabolic effect of a pituitary extract. Science *118:* 567.

Atkinson, D. E. (1968a). The energy charge of the adenylate pool as a regulatory parameter. Interaction with feedback modifiers. Biochemistry *7:* 4030–4034.

Atkinson, D. E. (1968b). Regulation of energy metabolism. Sympos. Fund. Cancer. Res. *22:* 397–413.

Atkinson, J. B. (1954). Factitial thyrotoxic crisis induced by dextro-amphetamine sulfate and thyroid. Ann. Intern. Med. *40:* 615–618.

Aurbach, G. D., Houston, B. A., and Potts, J. T., Jr. (1965). Control by parathyroid hormone of energy utilization in mitochondria. *In* Parathyroid Glands, Ultrastructure, Secretion, Function. pp. 197–205.

Azzi, A., Chance, B., Radda, G. K., and Lee, C. P. (1969). A fluorescence probe of energy-dependent structure changes in fragmented membranes. Proc. Nat. Acad. Sci. U.S.A. *62:* 612–619.

Bagdon, W. J., and Mann, D. E., Jr. (1965). Promazine hyperthermia in young albino mice. J. Pharm. Sci. *54:* 153–154.

Bain, J. A. (1954). The effect of magnesium upon thyroxine inhibition of phosphorylation. J. Pharmacol. Exp. Ther. *110:* 2–3.

Ball, E. G., and Cooper, O. (1957). The oxidation of reduced triphosphopyridine nucleotide as mediated by the transhydrogenase reaction, and its inhibition by thyroxine. Proc. Nat. Acad. Sci. U.S.A. *43:* 357–364.

Bansi, H. W. (1939). Die thyreotoxische Krise, das thyreotoxische Coma. Ergebn. Inn. Med. Kinderheilk. *56:* 305–371.

Barac, B. (1965). Effect of glucagon on the urinary excretion of inorganic phosphorus in the dog. C. R. Soc. Biol. (Paris) *159:* 1859–1861; CA *64:* 16241f (1966).

Barkan, I. N., and Mistakopulo, N. F. (1966). Action of some narcotics on oxidative and energy-forming processes in dog myocardial mitochondria. Dokl. Akad. Nauk. SSSR *173:* 683–686; CA *64:* 20432h (1966).

Barker, S. B. (1946). Effect of thyroid activity upon metabolic response to dinitro-ortho-cresol. Endocrinology *39:* 234–238.

Barker, S. B., Kiely, C. E., Jr., and Lipner, J. F. (1949). Metabolic effects of thyroxine injected into normal, thiouracil-treated and thyroidectomized rats. Endocrinology *45:* 624–626.

Barker, S. B., Pittman, C. S. Pittman, J. A., Jr., and Hill, S. R., Jr. (1960). Thyroxine antagonism by partially iodinated thyronines and analogues. Ann. N.Y. Acad. Sci. *86:* 545–562.

Bastenie, P. A., and Ermans, A. M. (1957). Effect de la cortisone sur l'action peripherique de la triiodothyronine. Helv. Med. Acta *24:* 188–192.

Bauman, T. R., and Turner, C. W. (1967). The effect of varying temperatures on thyroid activity and the survival of rats exposed to cold and treated with L-thyroxine or corticosterone. J. Endocr. *37:* 355–359.

Beaton, J. R. (1963). Phosphorus metabolism in cold exposed rats. Canad. J. Biochem. Physiol. *41:* 2209–2214.

Beattie, D. S., Basford, R. E., and Koritz, S. B. (1967). Bacterial contaminations and amino acid incorporation by isolated mitochondria. J. Biol. Chem. *242:* 3366–3368.

Bedrak, E. and Samoiloff, V. (1966). Aldosterone and oxidative phosphorylation in liver mitochondria. J. Endocr. *36:* 63–71.

Beechey, R. B., Alcock, N. W., and MacIntyre, I. (1961). Oxidative phosphorylation in magnesium and potassium deficiency in the rat. Amer. J. Physiol. *201:* 1120–1122.

Beierwaltes, W. H., Wolfson, W. Q., Jones, J. R., Knorpp, C. T., and Siemienski, J. S. (1950). Increase in basal oxygen consumption produced by cortisone in patients with untreated myxedema. J. Lab. Clin. Med. *36:* 799–800.

Belenky, M. L., and Vitolina, M. (1962). The pharmacological analysis of the hyperthermia caused by phenamine (amphetamine). Int. J. Neuropharmacol. *1:* 1–7.

Beyer, R. E. (1963). Regulation of energy metabolism during acclimation of laboratory rats to a cold environment. Fed. Proc. *22:* 874–880.

Beyer, R. E., Noble, W. M., and Hirschfeld, T. J. (1962). Alterations of rat-tissue coenyzme Q (ubiquinone) levels by various treatments. Biochim. Biophys. Acta *57:* 376-379.

Bing, R. J. (1961). Metabolic activity of the intact heart. Amer. J. Med. *30:* 679-691.

Blanchaer, M. C., and Wrogemann, K. (1968). Oxidative phosphorylation by mitochondria isolated from hearts of bio 14.6 myopathic hamsters. Trans. N.Y. Acad. Sci. *30:* 949-950.

Bodel, P., and Dillard, M. (1968). Studies on steroid fever. I. Production of leukocyte pyrogen *in vitro* by etiocholanone. J. Clin. Invest. *47:* 107–117.

Boothby, W. M., and Sandiford, I. (1922). Summary of the basal metabolism data on 8614 subjects with especial reference to the normal standards for the estimation of the basal metabolic rate. J. Biol. Chem. *54:* 783–803.

Borchardt, W. (1928). Fieber, Schilddrüse und Nebennieren. Klin. Wschr. *7:* 1507-1509.

Borison, H. L., and Clark, W. G. (1967). Drug actions on thermoregulatory mechanisms. Advances in Pharmacol. *5:* 129-212.

Borst, P., and Slater, E. C. (1960). The oxidation of glutamate by rat-heart sarcosomes. Biochim. Biophys. Acta *41:* 170-171.

Bosmann, H. B., and Martin, S. S. (1969). Mitochondrial autonomy: incorporation of monosaccharides into glycoprotein by isolated mitochondria. Science *164:* 190-192.

Boyer, P. D. (1964). Carboxyl activation as a possible common reaction in substrate-level and oxidative phosphorylation and in muscle contraction. In King, T. E., Mason, H. S., and Morrison, M. (eds.): Oxidases and Related Redox Systems. Vol. 2. New York, John Wiley & Sons, Inc., pp. 994–1008.

Brauer, R. W. (1963). Liver circulation and function. Physiol. Rev. *43:* 115–213.

Bray, G. A. (1969). Calorigenic effect of human growth hormone in obesity. J. Clin. Endocr. *29:* 119–122.

Bray, G. A., and Goodman, H. M. (1965). Studies on the early effects of thyroid hormones. Endocrinology *76:* 323–328.

Brenner, M., DeLorenzo, F., and Ames, B. N. (1970). Energy charge and protein synthesis. Control of aminoacyl transfer ribonucleic acid synthetases. J. Biol. Chem. *245:* 450–452.

Brewster, W. J., Jr., Isaacs, J. P., Osgood, P. F., and King, T. L. (1956). The hemodynamic and metabolic interrelationships in the activity of epinephrine, norepinephrine and the thyroid hormones. Circulation *13:* 1–20.

Brody, S. (1945). Bioenergetics and Growth. New York, Reinhold Publishing Corporation.

Brody, T. M. (1955). The uncoupling of oxidative phosphorylation as a mechanism of drug action. Pharmacol. Rev. *7:* 335–363.

Brody, T. M. (1956). Action of sodium salicylate and related compounds on tissue metabolism *in vitro*. J. Pharmacol. Exp. Ther. *117:* 39–51.

Brody, T. M., and Bain, J. A. (1951). Effect of barbiturates on oxidative phosphorylation. Proc. Soc. Exp. Biol. Med. *77:* 50–53.

Brody, T. M., and Bain, J. A. (1954). Barbiturates and oxidative phosphorylation. J. Pharmacol. Exp. Ther. *110:* 148–156.

Brody, T. M., and Killam, K. F. (1952). Potentiation of barbiturate anesthesia by 2,4-dinitrophenol. J. Pharmacol. Exp. Ther. (Proc.) *106:* 375.

Bronk, J. R. (1963). The nature of the energy requirement for amino acid incorporation by isolated mitochondria and its significance for thyroid hormone action. Proc. Nat. Acad. Sci. U.S.A. *50:* 524–526.

Bronk, J. R. (1966). Thyroid hormone: effects on electron transport. Science *153:* 638–639.

Bronk, J. R. (1968). Early effects of triiodothyronine *in vivo* on mitochondrial electron transport. Fed. Proc. *27:* 738.

Brown, D. M. (1966). Thyroxine stimulation of amino acid incorporation into protein of skeletal muscle *in vitro*. Endocrinology *78:* 1252–1254.

Brown-Grant, K. (1956). Changes in the thyroid activity of rats exposed to cold. J. Physiol. *131:* 52–57.

Brown-Grant, K., vonEuler, C., Harris, G. W., and Reichlin, S. (1954). The measurement and experimental modification of thyroid activity in the rabbit. J. Physiol. *126:* 1–28.

Buchanan, J., and Tapley, D. F. (1966). Stimulation by thyroxine of amino acid incorporation into mitochondria. Endocrinology *79:* 81–89.

Buchel, L., and McIlwain, H. (1950). Narcotics and the inorganic and creatine phosphates of mammalian brain. Brit. J. Pharmacol. *5:* 465–473.

Buffa, P., Carafoli, E., and Muscatello, U. (1960). Studies on the pathogenesis of hyperthermia produced by pentachlorophenol in the rat. Ital. J. Biochem. *9:* 248–253.

Buffa, P., Carafoli, E., and Muscatello, U. (1963). Mitochondrial biochemical lesion and pyrogenic effect of pentachlorophenol. Biochem. Pharmacol. *12:* 769–778.

Burgus, R., Dunn, T. F., Ward, D. N., Vale, W., Amoss, M., and Guiellemin, R. (1969). Dérivés polypeptidiques de synthèse doués d'activite hypophysiotrope TRF. C. R. Acad. Sci. (Paris) *268:* 2116–2118.

Burr, G. O., and Beber, A. J. (1937). Metabolism studies with rats suffering from fat deficiency. J. Nutr. *14:* 553–566.

Calvert, D. N., and Brody, T. M. (1961). The effects of thyroid function upon carbon tetrachloride hepatotoxicity. J. Pharmacol. Exp. Ther. *134:* 304–310.

Campbell, A. M., Corrance, M. H., Davidson, J. N., and Keir, H. M. (1969). The metabolism of DNA in the liver during precocious induction of metamorphosis in *Rana catesbiana*. Proc. Roy. Soc. (Biol.) (in press).

Carafoli, E. (1967). *In vivo* effect of uncoupling agents on the incorporation of calcium and strontium into mitochondria and other subcellular fractions of rat liver. J. Gen. Physiol. *50:* 1849–1864.

Carafoli, E., Margeth, A., and Buffa, P. (1963). Experiments on mitochondria from denervated muscle: effect of 4,6-dinitro-*o*-cresol *in vivo*. Ric. Sci. (Biol.) *3:* 255–260; CA *60:* 9498b (1964).

Carafoli, E., Rossi, C. S., and Lehninger, A. L. (1965). Energy-coupling in mitochondria during resting or State 4 respiration. Biochem. Biophys. Res. Commun. *19:* 609–614.

Carafoli, E., and Tiozzo, R. (1968). Energy-linked calcium transport in liver mitochondria during carbon tetrachloride intoxication. Exp. Molec. Path. *9:* 131–140.

Carlson, L. D. (1960). Nonshivering thermogenesis and its endocrine control. Fed. Proc. *19:* 25–30.

Cash, W. D., Aanning, H. L., Carlson, H. E., Cos, S. W., and Ekong, E. A. (1968). Role of zinc in the mitochondrial swelling action of insulin. Arch. Biochem. *128:* 456–459.

Cash, W. D., and Gardy, M. (1968). Failure of posterior pituitary hormone samples of low metal content to produce mitochondrial swelling. Endocrinology *83:* 368–370.

Cazeneuve, P., and Lépine, R. (1885). C. R. Acad. Sci. (Paris) *101:* 1167. Quoted in van Uytvanck, 1931.

Chaffee, R. R. J. (1962). Mitochondrial changes during the process of awakening from hibernation. Nature *196:* 789–780.

Chaffee, R. R. J., Hoch, F. L., and Lyman, C. (1961). Mitochondrial oxidative enzymes and phosphorylations in cold exposure and hibernation. Amer. J. Physiol. *201:* 29–32.

Chaffee, R. R. J., Horvath, S. M., Smith, R. E., and Welsh, R. S. (1966). Cellular biochemistry and organ mass of cold- and heat-acclimated monkeys. Fed. Proc. *25:* 1177–1184.

Challoner, D. R., and Steinberg, D. (1965). Metabolic effect of epinephrine on the Q_{O_2} of the arrested isolated perfused rat heart. Nature *205:* 602–603.

Chance, B. (ed.) (1963). Energy-linked functions of mitochondria. New York, Academic Press, Inc.

Chance, B. (1965). Reaction of oxygen with the respiratory chain in cells and tissues. J. Gen. Physiol. *49:* 163–195.

Chance, B., Bonner, W. D., Jr., and Storey, B. T. (1968a). Electron transport in respiration. Ann. Rev. Plant Physiol. *19:* 295–320.

Chance, B., Cohen, P., Jöbsis, F., and Schoener, B. (1962). Intracellular oxidation-reduction states *in vivo*. Science *137:* 499–508.

Chance, B., and Hess, B. (1959). Metabolic control mechanisms. I. Electron transfer in the mammalian cell. J. Biol. Chem. *234:* 2404–2412.

Chance, B., and Hollunger, G. (1963a). Inhibition of electron and energy transfer in mitochondria. I. Effects of Amytal, thiopental, rotenone, progesterone, and methylene glycol. J. Biol. Chem. *238:* 418–431.

Chance, B., and Hollunger, G. (1963b). Inhibition of electron and energy transfer in mitochondria. II. The site and the mechanism of guanidine action. J. Biol. Chem. *238:* 432–438.

Chance, B., and Hollunger, B. (1963c). Inhibition of electron and energy transfer in mitochondria. IV. Inhibition of energy-linked diphosphopyridine nucleotide reduction by uncoupling agents. J. Biol. Chem. *238:* 445–448.

Chance, B., Kohen, E., Kohen, C., and Legallais, V. (1967a). Microfluorimetric study of responses of the cytosol to mitochondrial substrates. Advances Enzym. Regulat. *5:*3–8.

Chance, B., Lee, C-P., and Mela, L. (1967b). Control and conservation of energy in the cytochrome chain. Fed. Proc. *26:* 1341–1354.

Chance, B., and Legallais, V. (1959). Differential microfluorimeter for the localization of reduced pyridine nucleotides in living cells. Rev. Sci. Instrum. *30:* 732–735.

Chance, B., Mela, L., and Harris, E. J. (1968b). Interaction of ion movements and local anesthetics in mitochondrial membranes. Fed. Proc. *27:* 902–906.

Chance, B., Perry, R., Akerman, L., and Thorell, B. (1959). Highly sensitive recording microspectrophotometer. Rev. Sci. Instrum. *30:* 735–741.

Chance, B., Williams, G. R., and Hollunger, G. (1963). Inhibition of electron and energy transfer in mitochondria. III. Spectroscopic and respiratory effects of uncoupling agents. J. Biol. Chem. *238:* 439–444.

Chance, B., Williamson, J. R., Jamieson, D., and Schoener, B. (1965). Properties and kinetics of reduced pyridine nucleotide fluorescence of the isolated and *in vivo* rat heart. Biochem. Z. *341:* 357–377.

Charnock, J. S., Lockett, R., and Hetzel, B. S. (1962). Effect of salicylate on plasma magnesium levels. Nature *195:* 295–296.

Chase, L. R., Fedak, S. A., and Aurbach, G. D. (1969). Activation of skeletal adenyl cyclase by parathyroid hormone *in vitro*. Endocrinology *84:* 761–768.

Christiansen, E. N., Pedersen, J. I., and Grav, H. J. (1969). Uncoupling and recoupling of oxidative phosphorylation in brown adipose tissue mitochondria. Nature *222:* 857–860.

Chu, H., Opitz, K., and Intemann, E. (1969). Role of the thyroid in the calorigenic action of amphetamine. Naunyn Schmiedeberg Arch. Exp. Path. *263:* 358–362.

Clark, J. H., Jr., and Pesch, L. A. (1956). Effects of cortisone upon liver enzymes and protein synthesis. J. Pharmacol. Exp. Ther. *117:* 202–207.

Climenko, D. R. (1936). Influence of magnesium oxide on antipyretic action and toxicity of acetylsalicylic acid in rabbits. Proc. Soc. Exp. Biol. Med. *34:* 807–812.

Cochran, J. B. (1952). The respiratory effects of salicylate. Brit. Med. J. *2:* 964–967.

Cochran, J. B. (1954). Further observations on the metabolic stimulating effect of salicylate. Brit. Med. J. *1:* 733–743.

Coffey, D. S., Ichinose, R. R., Shimazaki, J., and Williams-Ashman, H. B. (1968). Effects of testosterone on adenosine triphosphate and nicotinamide adenine dinucleotide levels and on nicotinamide mononucleotide adenyltransferase activity in the ventral prostate of castrated rats. Molec. Pharmacol. *4:* 580–590.

Coh, D. V., Smaich, A. F., and Levy, R. (1966). The inhibition of respiration and phosphorylation in kidney mitochondria by parathyroid hormone administered *in vivo*. J. Biol. Chem. *241:* 889–894.

Cohn, D. V., Bawdon, R. R., and Hamilton, J. W. (1968). Effect of calcium chelation on the ion content of liver mitochondria in carbon tetrachloride-poisoned rats. J. Biol. Chem. *243:* 1099–1095.

Colowick, S. P., Kaplan, N. O., Neufeld, E. F., and Ciotti, M. M. (1952). Pyridine nucleotide transhydrogenase. I. Indirect evidence for the reaction and purification of the enzyme. J. Biol. Chem. *195:* 95–105.

Colowick, S. P., van Eys, J., and Park, J. H. (1966). Dehydrogenation. *In* Florkin, M., and Stotz, E. H. (eds.): Comprehensive Biochemistry. Vol. 14. New York, American Elsevier Publishing Company, Inc., pp. 1–98.

Conney, A. H. (1969). Drug metabolism and therapeutics. New Eng. J. Med. *280:* 653–660.

Conney, A. H., and Garren, L. (1961). Contrasting effects of thyroxine on zoxazolamine and hexobarbital metabolism. Biochem. Pharmacol. *6:* 257–262.

Contopoulos, A. N., Evans, E. E., Ellis, S., and Simpson, M. E. (1954). Increased metabolic rate without thyroid participation on injection of rats with pituitary erythropoietic fractions. Proc. Soc. Exp. Biol. Med. *86:* 729–733.

Corradino, R. A., and Parker, H. E. (1962). Magnesium and thyroid function in the rat. J. Nutr. *77:* 455–458.

Correspondent, Medical Biochemistry (1968). Mitochondria producing heat. Nature 218: 321–322.

Corssen, G., Sweet, R. B., and Chenoweth, M. B. (1966). Effects of chloroform, halothane and methoxyflurane on human liver cells in vitro. Anesthesiology 27: 155–162.

Crabtree, B., and Newsholme, E. A. (1969). Importance of α-glycerophosphate oxidase in oxidation of extramitochondrial reduced nicotinamide adenine dinucleotide in vertebrate and insect muscle. Biochem. J. 114: 80P–81P.

Crane, F. L., and Löw, H. (1966). Quinones in energy-coupling systems. Physiol. Rev. 46: 662–695.

Crawford, A. L., Henderson, M. J., Hawkins, R. D., and Haist, R. E. (1965). The effect of glucagon on blood sugar and inorganic phosphorus levels in normothermic and hypothermic rats. Canad. J. Physiol. Pharmacol. 43: 601–610.

Cross, R. J., Taggert, J. V., Covo, G. A., and Green, D. E. (1949). Studies on the cyclophorase system. VI. The coupling of oxidation and phosphorylation. J. Biol. Chem. 177: 655–678.

Currie, W. D., and Gregg, C. T. (1965). Inhibition of the respiration of cultured mammalian cells by oligomycin. Biochem. Biophys. Res. Commun. 21: 9–15.

Cutting, W. C., Mehrtens, H. G., and Tainter, M. L. (1933). Actions and uses of dinitrophenol. J.A.M.A. 101: 193–195.

Dallner, G., and Ernster, L. (1962). Induction of a Crabtree-like effect in Ehrlich ascites tumor cells by oligomycin. Exp. Cell Res. 27: 372–375.

Davidson, I. W. F., Salter, J. M., and Best, C. H. (1960). The effect of glucagon on the metabolic rate of rats. Amer. J. Clin. Nutr. 8: 540–546.

Dawson, J., and Jones, E. A. (1957). The in vitro effect of parathormone on oxidative phosphorylation. Scand. J. Clin. Lab. Invest. (Suppl.) 31: 264–265.

deHaan, E. J., Tager, J. M., and Slater, E. C. (1967). Factors affecting the pathway of glutamate oxidation in rat liver mitochondria. Biochim. Biophys. Acta 131: 1–13.

DeLuca, H. F., Engstrom, G. W., and Rasmussen, H. (1962). The action of vitamin D and parathyroid hormone in vitro on calcium uptake and release by kidney mitochondria. Proc. Nat. Acad. Sci. U.S.A. 48: 1604–1609.

DeMeio, R. H., and Barron, E. S. G. (1934–35). Effects of 1,2,4-dinitrophenol on cellular respiration. Proc. Soc. Exp. Biol. Med. 32: 36–39.

Denborough, M. A., Forster, F. A., Lovell, R. H., Maplestone, P. A., and Villiers, J. D. (1962). Anaesthetic deaths in a family. Brit. J. Anaesth. 34: 395–396.

Denis, W., and Means, J. H. (1916). The influence of salicylate on metabolism in man. J. Pharmacol. Exp. Ther. 8: 273–283.

Depocas, F. (1961). Biochemical changes in exposure and acclimation to cold environments. Brit. Med. Bull. 17: 25–31.

Derache, R., Tremolières, J., Griffaton, G., and Lowy, R. (1957). Effects de la cortisone sur la synthese aérobie de l'acide adénosine triphosphorique (ATP). Bull. Soc. Chim. Biol. 39: 607–618.

Derrick, J. B., and Collip, J. B. (1953). Further studies on a pituitary factor with specific influence on metabolism. Canad. J. Med. Sci. 31: 117–125.

Des Marais, A., and LeMay, B. (1964). Effects of an inhibitor of the calorigenic activity of thyroxine in warm- and cold-acclimated mice. Canad. J. Physiol. Pharmacol. 42: 373–376.

Devlin, T. M. (1959). Respiratory chain in phosphorylating subfragments of mitochondria prepared with digitonin. J. Biol. Chem. 234: 962–966.

Dianzani, M. U. (1954). Uncoupling of oxidative phosphorylation in mitochondria from fatty livers. Biochim. Biophys. Acta 14: 514–532.

Dianzani, M. U., and Dianzani-Mor, M. A. (1957). Decrease of phosphorus/oxygen ratios by dinitrophenol in vivo. Nature 179: 532.

Dianzani, M. U., and Scuro, S. (1956). The effects of some inhibitors of oxidative phosphorylation on the morphology and enzymic activities of mitochondria. Biochem. J. 62: 205–215.

Dickerson, R. E., Kopka, M. L., Weinzierl, J. E., Varnum, J. C., Eisenberg, D., and Margoliash, E. (1968). An interpretation of a two-derivative, 4 A° resolution electron density map of horse heart ferricytochrome c. In Okunuki, K., Kamen, M. D., and Sekuzu, I. (eds): Function of Cytochromes, Baltimore, University Park Press, pp. 225–251.

Dietrich, W. C., and Beutner, R. (1944). The mechanism of action of thiouracil. Proc. Soc. Exp. Biol. Med. *57:* 35-36.

Dillon, R. S., and Hoch, F. L. (1967). Iodine in mitochondria and nuclei. Biochem. Med. *1:* 219–229.

Dingman, C. W., and Sporn, M. B. (1965). Actinomycin D and hydrocortisone: intracellular binding in rat liver. Science *149:* 1251–1254.

Dodd, K., Minot, A. S., and Arena, J. M. (1937). Salicylate poisoning: an explanation of the more serious consequences. Amer. J. Dis. Child. *53:* 1435–1446.

Dodds, E. C., and Robertson, J. D. (1933). The clinical applications of dinitro-*o*-cresol. Lancet *2:* 1137–1139, 1197–1198.

Donhoffer, S., Balogh, L., and Mestyan, G. (1949). The immediate response of thyroidectomized rats to small doses of thyroxine. Experientia *5:* 482–483.

Donhoffer, S., Varnai, I., and Sziebert-Horvath, E. (1958). Immediate effect of L-3 5,3'-triiodothyroacetic acid on metabolic rate and body temperature in hypophysectomized rats and the action of cortisone. Nature *181:* 345–346.

Dow, D. S. (1967). The isolation of skeletal muscle mitochondria showing tight-coupling, high respiratory indices, and differential adenosine triphosphatase activities. Biochemistry *6:* 2915–2922.

Drabkin, D. G. (1950). Cytochrome *c* metabolism and liver regeneration. Influence of thyroid gland and thyroxine. J. Biol. Chem. *182:* 335–349.

Dryer, R. L., and Paulsrud, J. R. (1966). Effect of arousal on ATP levels in bats. Fed. Proc. *25:* 1293–1296.

DuBois, E. F. (1916). Clinical calorimetry. XIV. Metabolism in exophthalmic goiter. Arch. Intern. Med. *17:* 915–964.

DuBois, E. F. (1936). Basal Metabolism in Health and Disease. Philadelphia, Lea and Febiger.

Dunlop, D. M. (1934). The use of 2,4-dinitrophenol as a metabolic stimulant. Brit. Med. J. *1:* 524–527.

DuToit, C. (1951). The effects of thyroxine on phosphate metabolism. *In* McElroy, W. D., and Glass, B. (eds.): Phosphorus metabolism. Vol. I. Baltimore, Johns Hopkins Press, pp. 597–617.

Edelman, I. S., and Fimognari, G. M. (1968). On the biochemical mechanism of action of aldosterone. Recent Progr. Hormone Res. *24:* 1–34.

Editorial (1966). Malignant hyperpyrexia during general anaesthesia. Canad. Anaesth. Soc. J. *13:* 415–416.

Ehrenfest, E., and Ronzoni, E. (1933). Effect of dinitrophenol on oxidation of tissues. Proc. Soc. Exp. Biol. Med. *31:* 318–319.

Ernster, L., Ikkos, D., and Luft, R. (1959). Enzymic activities of human skeletal muscle mitochondria: a tool in clinical metabolic research. Nature *184:* 1851–1854.

Ernster, L., Lee, C.-P., and Janda, S. (1967). The reaction sequence in oxidative phosphorylation. *In* Slater, E. C., Kaniuga, Z., and Wojtczak, L. (eds.): Biochemistry of Mitochondria. New York, Academic Press, Inc., pp. 29–52.

Ernster, L., and Luft, R. (1963). Population of human skeletal muscle mitochondria lacking respiratory control. Exp. Cell Res. *32:* 26–35.

Ernster, L., and Luft, R. (1964). Mitochondrial respiratory control: biochemical, physiological, and pathological aspects. Advances Metab. Dis. *1:* 95–123.

Evans, E. S., Contopoulos, A. N., and Simpson, M. E. (1957). Hormonal factors influencing calorigenesis. Endocrinology *60:* 403–419.

Evans, E. S., Simpson, M. E., and Evans, H. M. (1958). The role of growth hormone in calorigenesis and thyroid function. Endocrinology *63:* 836–852.

Falcone, A. B. (1959). Studies on the mechanism of action of salicylates; effects on oxidative phosphorylation. J. Clin. Invest. (Abstracts) *38:* 1002-1003.

Feinstein, M. B. (1964). Reaction of local anesthetics with phospholipids. A possible chemical basis for anesthesia. J. Gen. Physiol. *48:* 357–374.

Field, J., 2nd, Belding, H. S., and Martin, A. W. (1939). An analysis of the relation between basal metabolism and summated tissue respiration in the rat. J. Cell. Comp. Physiol. *14:* 143–157.

Firkin, F. C., and Linnane, A. W. (1969). Biogenesis of mitochondria. VIII. Effect of chloramphenicol on regenerating rat liver. Exp. Cell Res. *55:* 68–76.

Fisher, R. B., and Williamson, J. R. (1961). The effects of insulin, adrenaline, and nutrients on the O uptake of the perfused rat heart. J. Physiol. *158:* 102–112.

Flock, E. V., and Bollman, J. L. (1964). Effect of butyl 4-hydroxy-3,5-diiodo-benzoate on the metabolism of L-thyroxine and related compounds. Endocrinology *75:* 721–732.

Florijn, E., Gruber, M., Leijnse, B., and Huisman, T. H. J. (1950). The inhibition of tissue respiration and alcoholic fermentation at different catabolic levels by ethyl carbamate (urethane) and arsenite. Biochim. Biophys. Acta *5:* 595–605.

Foa, P. P., and Galansino, G. (1962). Glucagon: chemistry and function in health and disease. Springfield, Ill, Charles C Thomas, Publisher.

Forester, C. F. (1963). Coma in myexdema: report of a case and review of the world literature. Arch. Intern. Med. *111:* 734-743.

Frank, G. B., and Sanders, H. D. (1963). A proposed common mechanism of action for general and local anaesthetics in the central nervous system. Brit. J. Pharmacol. *21:* 1–9.

Fraser, T. R., and Maclagan, N. F. (1953). Clinical trial of an antithyroxine compound. J. Endocr. *9:* 301–306.

Frawley, T. F., and Zacharewicz, F. A. (1962). Antimetabolic activity of a thyroxine analogue: 3,5-diiodothyroacetic acid (DIAC). Metabolism *11:* 579–588.

Freedland, R. A., Avery, E. H., and Taylor, A. R. (1968). Effect of thyroid hormones on metabolism. II. The effect of adrenalectomy or hypophysectomy on responses of rat liver enzyme activity to L-thyroxine injection. Canad. J. Biochem *46:* 141-150.

Freeman, K. B., Roodyn, D. B., and Tata, J. R. (1963). Stimulation of amino acid incorporation into protein by isolated mitochondria from rats treated with thyroid hormones. Biochim. Biophys. Acta *72:* 129–132.

Freeman, M. E., Crissman, J. K., Jr., Louw, G. N., Butcher, R. L., and Inskeep, E. K. (1970). Thermogenic action of progesterone in the rat. Endocrinology *86:* 717–720.

Fregly, M. J. (1967). Metabolic responses to heat. Proc. Third Midwest Conf. on the Thyroid, 1967. Columbia, University of Missouri, p. 1–20.

Friedenwald, J. S., and Buschke, W. (1943). The effect of cyanide and other metal-binding substances on the pharmacological action of adrenaline. Amer. J. Physiol. *140:* 367–373.

Gaebler, O. H. (1933). Some effects of anterior pituitary extracts on nitrogen metabolism, water balance, and energy metabolism. J. Exp. Med. *57:* 349–363.

Gambal, D., and Quackenbush, F. W. (1968). Essential fatty acids, plasma protein bound iodine, and the thyroid gland. Proc. Soc. Exp. Biol. Med. *127:* 1137–1138.

Gambetti, P., Gonatas, N. K., and Flexner, L. B. (1968). Puromycin action on neuronal mitochondria. Science *161:* 900–902.

Ganju, S. N., and Lockett, M. F. (1958). The action of thyroid hormones in the oxygen consumption and resistance to cold of adrenalectomized and thyroidectomized mice. J. Endocr. *16:* 346–402.

Gemmill, C. L., and Browning, K. M. (1962). Effects of pentobarbital on temperature and heart rate of rats subjected to cold. Amer. J. Physiol. *203:* 758–761.

Gemmill, C. L., Browning, K. M., and Gemmill, D. D. (1962). Effects of iodinated salicylates on metabolism of rats and mice. Proc. Soc. Exp. Biol. Med. *110:* 39–41.

George, P., and Rutman, R. J. (1960). The high-energy phosphate bond concept. Progr. Biophys. *10:* 1–53.

Gertler, M. M. (1961). Differences in efficiency of energy transfer in mitochondrial systems derived from normal and failing hearts. Proc. Soc. Exp. Biol. Med. *106:* 109–112.

Gessa, G. L., and Clay, G. A. (1969). Evidence that hyperthermia produced by *d*-amphetamine is caused by a peripheral action of the drug. Life Sci. *8:* 135–141.

Giacovazzo, M., Bianchi, P., and LaTorre, F. (1959). ATP and glucagon; variations in the ATP content of serum, liver, and muscle after treatment with glucagon. Rass. Fisiopat. Clin. Ter. *31:* 934–943; CA *62:* 793b (1965).

Glaubach, S., and Pick, E. P. (1930). Über die Beeinflussung der Temperaturregulierung durch Thyroxin. I. Mitteilung. Arch. Exp. Path. Pharmakol. *151:* 341-370.

Glaubach, S., and Pick, E. P. (1931). Über die Beeinflussung der Temperaturregulierung durch Thyroxin. II. Kokain-, Perkain- und Novokainwirkung bei thyroxinvorbehandelten Tieren. Arch Exp. Path. Pharmakol. *162:* 537–550.

Glaubach, S., and Pick, E. P. (1934). Über den Einflusz der Schilddrüse auf die Arzneiempfindlichkeit. Schweiz. Med. Wschr. *15:* 1115-1116.

Goldhaber, P. (1963). Some chemical factors influencing bone resorption in tissue culture. *In* Sognnaes, R. F. (ed.): Mechanisms of Hard Tissue Destruction, Phila., 1962. Amer. Ass. Adv. Sci., pp. 609–636.

Goodman, L. S., and Gilman, A. (eds.) (1967). The Pharmacological Basis of Therapeutics. The 3rd ed. New York, Macmillan Company.

Gorski, J., Toft, D., Shyamala, G., Smith, D., and Notides, A. (1968). Hormone receptors: studies on the interaction of estrogen with the uterus. Recent Progr. Hormone Res. *24:* 45–80.

Green, D. E. (1941). Enzymes and trace substances. Advances Enzym. *1:* 177–198.

Green, D. E. (1966). The mitochondrial electron-transfer system. *In* Florkin, M., and Stotz, E. (eds.): Comprehensive Biochemistry. Vol. 14. New York. Academic Press, Inc., pp. 309–326.

Green, D. E., Asai, J., Harris, R. A., and Penniston, J. T. (1968). Conformational basis of energy transformations in membrane systems. III. Configurational changes in the mitochondrial inner membrane induced by changes in functional states. Arch. Biochem. *125:* 684–705.

Green, D. E., and Baum. H. (1970). Energy and the Mitochondrion.. New York, Academic Press, Inc.

Gregg, C. T., Machinist, J. M., and Currie, W. D. (1968). Glycolytic and respiratory properties of intact mammalian cells: inhibitor studies. Arch. Biochem. *127:* 101–111.

Greville, G. D. (1969). A scrutiny of Mitchell's chemiosmotic hypothesis of respiratory chain and photosynthetic phosphorylation. Curr. Topics in Bioenergetics *3:* 1–78.

Griffith, F. R., Jr. (1951). Fact and theory regarding the calorigenic action of adrenaline. Physiol. Rev. *31:* 151–187.

Guillory, R. J. (1969). Action of uncouplers and antibiotics in oxidative phosphorylation: prospectives for π-interaction. Ann. N.Y. Acad. Sci. *153:* 815–825.

Guillory, R. J., and Racker, E. (1968). Oxidative phosphorylation in brown adipose mitochondria. Biochim. Biophys. Acta *153:* 490–493.

Gyermek, L. (1950). Die Wirkung des Aktedrons (Phenylisopropylamins) auf den Gaswechsel an nebennierenlosen und mit Thyroxin behandelten Tieren. Arch. Exp. Path. Pharmakol. *209:* 27–34.

Hackenbrock, C. R. (1966). Ultrastructural bases for metabolically linked mechanical activity in mitochondria. I. Reversible ultrastructural changes with change in metabolic steady state in isolated liver mitochondria. J. Cell Biol. *30:* 269–297.

Hackenbrock, C. R. (1968). Chemical and physical fixation of isolated mitochondria in low-energy and high-energy states. Proc. Nat. Acad. Sci. U.S.A. *61:* 598–605.

Hall, J. C., Sordhal, L. A., and Stefko, P. L. (1960). The effect of insulin on oxidative phosphorylation in normal and diabetic mitochondria. J. Biol. Chem. *235*: 1536–1539.

Hall, L. W., Woolf, N., Bradley, J. W. P., and Jolly, D. W. (1966). Unusual reaction to suxamethonium chloride. Brit. Med. J. *2:* 1305.

Halmi, N. S., Nissen, Wm. M., and Scranton, J. R. (1969). Kinetics of the enhancement of thyroidal iodide accumulation after actinomycin D administration. Endocrinology *84:* 943–945.

Hannon, J. P. (1960). Tissue energy metabolism in the cold-acclimatized rat. Fed. Proc. *19:* 139–144.

Hannon, J. P., and Rosenthal, A. (1963). Effects of cold acclimatization on liver di- and triphosphopyridine nucleotide. Amer. J. Physiol. *204:* 515–516.

Harary, I., and Slater, E. C. (1965). Studies *in vitro* on single beating heart cells. VIII. The effect of oligomycin, dinitrophenol and ouabain on the beating rate. Biochim. Biophys. Acta *99:* 227–233.

Hardy, J. D. (1961). Physiology of temperature regulation. Physiol. Rev. *41:* 521–606.

Harris, R. A., Harris, D. L., and Green, D. E. (1968a). Effect of Bordetella endotoxin upon mitochondrial respiration and energized processes. Arch. Biochem. *128:* 219–230.

Harris, R. A., Penniston, J. T., Asai, J., and Green, D. E. (1968b). The conformational basis of energy conservation in membrane systems. II. Correlation between conformational change and functional states. Proc. Nat. Acad. Sci. U.S.A. *59:* 830-837.

Harrison, G. G., Biebuyck, J. F., Terblanche, J., Dent, D. M., Hichman, R., and Saunders, S. J. (1968). Hyperpyrexia during anaesthesia. Brit. Med. J. *3:* 594–595.

Harrison, M. T. (1968). The prevention and treatment of thyroid storm. Pharmacol. for Physicians *1:* 2.

Harrison, T. S. (1964). Adrenal medullary and thyroid relationships. Physiol. Rev. *44:* 161–185.

Hart, J. S. (1969). Proc. Int. Sympos. on Altitude and Cold, 1968. Fed. Proc. *28:* 933–1321.

Hayashida, T., and Portman, O. W. (1963). Changes in succinic dehydrogenase activity and fatty acid composition of rat liver mitochondria in essential fatty acid deficiency. J. Nutr. *81:* 103–109.

Hemingway, A., Price, W. M., and Stuart, D. (1964). The calorigenic action of catecholamines in warm acclimated and cold acclimated non-shivering cats. Int. J. Neuropharmacol. *3:* 495–503.

Henley, K. S. (1962). The effect of bilateral adrenalectomy on the oxidation of alphaglycerophosphate (AGP) in rat liver mitochondria. Aktuelle Probleme der Hepatologie. II. Sympos. Int. Ass. Study Liver, 1962. Stuttgart, Germany, Georg Thieme Verlag KG., pp. 195–198.

Henley, K. S., Kawata, H., and Pino, M. E. (1963). Effect of adrenalectomy on rat liver mitochondria. Endocrinology *73:* 366–370.

Hepp, D., Challoner, D. R., and Williams, R. H. (1968). Respiration in isolated fat cells and the effects of epinephrine. J. Biol. Chem. *243:* 2321–3227.

Hervey, G. R. (1969). Regulation of energy balance. Nature *222:* 629–631.

Hess, B., and Brand, K. (1964). Wirkungsmechanismus des Schilddrüsenhormons. Sympos. der Deut. Gesell. für Endokrin., Vienna, 1963. Berlin-Göttingen-Heidelberg, Springer-Verlag OHG.

Heymans, C., and Bouckaert, J. J. (1932). Hyperthermia by dinitro-α-naphthol. Acta Brev. Neerland. Physiol. Pharmacol. Microbiology *2:* 99; CA *26:* 6019.

Heymans, C., and Casier, H. (1933). Action stimulante, sur le métabolisme cellulaire, du dinitrocrésol et du dinitrothymol. C. R. Soc. Biol. (Paris) *114:* 1384–1385.

Heymans, C., and Reginers, P. (1926). Influence de quelques anesthésiques et hypnotiques sur l'hyperthermie par le bleu de méthylène. Arch. Int. Pharmacodyn. *32:* 311–326.

Hillmann, G. (1961). Biosynthese und Stoffwechselwirkungen der Schilddrüsenhormone. Stuttgart, Germany, Georg Thieme Verlag KG.

Hoch, F. L. (1962a). Biochemical actions of thyroid hormones. Physiol. Rev. *42:* 605–673.

Hoch, F. L. (1962b). Thyrotoxicosis as a disease of mitochondria. New Eng. J. Med. *266:* 446–454, 498–505.

Hoch, F. L. (1965a). Synergism between calorigenic effects: L-thyroxine and 2,4-dinitrophenol or sodium salicylate in euthyroid rats. Endocrinology *76:* 335–339.

Hoch, F. L. (1965b). L-Thyroxine in subcalorigenic doses: rapid potentiation of dinitrophenol-induced calorigenesis in hypothyroid rats. Endocrinology *77:* 991–998.

Hoch, F. L. (1967a). Biochemical action of thyroid hormones. *In* Proc. Third Midwest Conf. on Thyroid, University of Missouri, 1967. University of Missouri Press, pp. 89–107.

Hoch, F. L. (1967b). Early action of injected L-thyroxine on mitochondrial oxidative phosphorylation. Proc. Nat. Acad. Sci. U.S.A. *58:* 506–512.

Hoch, F. L. (1968a). Thyroid hormone action on mitochondria. I. Effects of inhibitors of respiration. Arch. Biochem. *124:* 238–247.

Hoch, F. L. (1968b). Thyroid hormone action on mitochondria. II. Effects of dinitrophenol. Arch. Biochem. *124:* 248–257.

Hoch, F. L. (1968c). Biochemistry of hyperthyroidism and hypothyroidism. Postgrad. Med. J. *44:* 347–362.

Hoch, F. L. (1970). The pharmacologic basis for the clinical use of thyroid hormones, Pharmacol. for Physicians. *4:* 1–5.

Hoch, F. L., and Cahill, G. F., Jr. (1966). Intermediary metabolism. *In* Harrison, T. R., *et al.* (eds.): Principles of Internal Medicine. 5th ed. New York, McGraw-Hill Book Co., pp. 333–344.

Hoch, F. L., and Lipmann, F. (1954). The uncoupling of respiration and phosphorylation by thyroid hormones. Proc. Nat. Acad. Sci. U.S.A. *40:* 909–921.

Hoch, F. L., and Motta, M. V. (1968). Reversal of early thyroid hormone action on mitochondria by bovine serum albumin *in vitro.* Proc. Nat. Acad. Sci. U.S.A. *59:* 118–122.

Hoffmann, F., Hoffmann, E. J., and Talesnik, J. (1948). The influence of thyroxine and adrenal cortical extract on the oxygen consumption of adrenalectomized rats. J. Physiol. _107:_ 251–264.

Hofmann-Credner, D., and Siedek, H. (1949). Über den Einflusz von Vitamin A und von Methylthiouracil auf die Dinitro-_o_-kresolwirkung bei Mensch und Tier. Klin. Med. _4:_ 361–367.

Hohorst, J.-J., Stratmann, D., and Bartels, H. (1963). The effect of insulin on the products and enzymes of energy metabolism in the liver. Deutsch. Med. Forsch. _1:_ 199–201.

Holloszy, J. O., and Oscai, L. B. (1969). Effect of exercise on α-glycerophosphate dehydrogenase activity in skeletal muscle. Arch. Biochem. _130:_ 653–656.

Holloway, C. T., Bond, R. P. M., Knight, I. G., and Beechey, R. B. (1967). The alleged presence and role of monoiodohistidine in mitochondrial oxidative phosphorylation. Biochemistry _6:_ 19–25.

Holloway, S. A., and Stevenson, J. A. F. (1964). Effect of glucagon on oxygen consumption in the cold-adapted rat. Canad. J. Physiol. Pharmacol. _42:_ 860–862.

Holtkamp, D. E., and Heming, A. E. (1953). Prevention of thyroxin-induced increased oxygen consumption of rats by current "Dibenzyline" administration. Fed. Proc. _12:_ 331.

Hommes, F., and Estabrook, R. W. (1963). The role of transhydrogenase in the energy-linked reduction of TPN. Biochem. Biophys. Res. Commun. _11:_ 1–6.

Horwitz, B. A., Herd, P. A., and Smith, R. E. (1968). Effect of norepinephrine and uncoupling agents on brown adipose tissue. Canad. J. Physiol. Pharmacol. _46:_ 897–902.

Hueber, E. F. V. (1939). Über die Beeinflussung von Hyperthyreosen durch Magnesiumglutaminat. Wien. Klin. Wschr. _52:_ 932–933.

Hulsmann, W. C., Bethlem, J., Meijer, A. E. F. H., Fleury, P., and Schellens, J. P. M. (1967). Myopathy with abnormal structure and function of muscle mitochondria. J. Neurol. Neurosurg. Psychiat. _30:_ 519–525.

Hunter, F. E., Jr., and Ford, L. (1955). Inactivation of oxidative and phosphorylative systems in mitochondria by preincubation with phosphate and other ions. J. Biol. Chem. _216:_ 357–369.

Hurwitz, L. J., McCormick, D., and Allen, I. V. (1970). Reduced muscle α-glucosidase (acid-maltase) activity in hypothyroid myopathy. Lancet _1:_ 67–68.

Huston, M. J., and Martin, A. W. (1954). Rate of respiration of tissues in contact with oxygen. Proc. Soc. Exp. Biol. Med. _86:_ 103–107.

Ichihara, A., Tanioka, H., and Takeda, Y. (1965). Respiratory control of dispersed rat-liver cells. Biochim. Biophys. Acta _97:_ 1–8.

Ikemoto, H., Hiroshige, T., and Itoh, S. (1967). Oxygen consumption of brown adipose tissue in normal and hypothyroid mice. Jap. J. Physiol. _17:_ 516–522.

Isaacs, G. H., Sacktor, B., and Murphy, T. A. (1969). Role of the α-glycerophosphate cycle in the control of carbohydrate oxidation in heart and in the mechanism of action of thyroid hormone. Biochim. Biophys. Acta _177:_ 196–203.

Ito, T., and Johnson, R. M. (1964). Effects of a nutritional deficiency of unsaturated fats on rat liver mitochondria. I. Respiratory control and adenosine triphosphate-inorganic orthophosphate exchange activity. J. Biol. Chem. _239:_ 3201–3203.

Itoh, S., Hiroshige, T., Koseki, T., and Nakatsugawa, T. (1966). Release of thyrotopin in relation to cold exposure. Fed. Proc. _25:_ 1187–1194.

Jarrett, L. (1967). Evidence for the use of high energy intermediate(s) for protein synthesis in isolated fat cells. J. Lab. Clin. Med. (Proc. Cent. Soc. Clin. Res.) _70:_ 1021.

Jarrett, L., and Kipnis, D. M. (1967). Differential response of protein synthesis in Ehrlich ascites tumor cells and normal thymocytes to 2,4-dinitrophenol and oligomycin. Nature _216:_ 714–715.

Jencks, W. P. (1963). The chemistry of biological energy transfer. Survey Progr. Chem. _1:_ 249–300.

Johnson, P. C., Posey, A. F., Patrick, D. R., and Caputto, R. (1958). Incorporation of P_{32} in the muscle by normal and thyrotoxic resting rats. Amer. J. Physiol. _192:_279–282.

Johnson, R. M. (1963). Adenosine triphosphatase and ATP-P_i exchange in mitochondria of essential fatty acid-deficient rats. J. Nutr. _81:_ 411–414.

Johnston, C. C., and Bartlett, P. (1965). Respiratory control in kidney and liver mitochondria isolated from rats treated with the potent nephrotogenic aminonucleoside of puromycin. Biochem. Pharmacol. _14:_ 1231–1236.

Jolly, W., Harris, R. A., Asai, J., Lenaz, G., and Green, D. E. (1969). Studies of ultrastructural dislocations in mitochondria. II. On the dislocation induced by lyophilization and the mechanism of uncoupling. Arch. Biochem. *130:* 191–211.

Jones, C. T., and Banks, P. (1969). Inhibition of respiration by puromycin in slices of cerebral cortex. J. Neurochem. *16:* 825–828.

Jones, J. E., Desper, P. C., Shane, S. R., and Flink, E. B. (1966). Magnesium metabolism in hyperthyroidism and hypothyroidism. J. Clin. Invest. *45:* 891–900.

Kaciuba-Uscilko, H., Legge, K. F., and Mount, L. E. (1970). The development of the metabolic response to thyroxine in the new-born pig. J. Physiol. *206:* 229–241.

Kadenbach, B. (1966). Der Einflusz von Thyreiodhormonen *in vivo* auf die oxidative Phosphorylierung und Enzymaktivitäten in Mitochondrien. Biochem. Z. *344:* 49–75.

Kadis, S., Montie, T. C., and Ajl, S. J. (1969). Plague toxin. Sci. Amer. *220:* 93–100.

Kalant, N., and Clamen, M. (1959). Dietary magnesium and oxygen consumption. Arch. Biochem. *85:* 563–564.

Kalckar, H. M. (1969). Biological Phosphorylation: Development of Concepts. Englewood Cliffs, New Jersey, Prentice-Hall, Inc.

Kaldor, G. (1969). Physiological Chemistry of Proteins and Nucleic Acids in Mammals. (Physiological Chemistry Series, E. J. Masoro, ed.). Philadelphia, W. B. Saunders Company.

Kalf, G. F. (1964). Deoxyribonucleic acid in mitochondria and its role in protein synthesis. Biochem. *3:* 1702–1706.

Kaplan, N. O., Colowick, S. P., and Neufeld, E. F. (1953). Pyridine nucleotide transhydrogenase. III. Animal tissue transhydrogenases. J. Biol. Chem. *205:* 1–16.

Kaplan, N. O., Swartz, M. N., French, M. E., and Ciotti, M. M. (1956). Phosphorylative and nonphosphorylative pathways of electron transfer in rat liver mitochondria. Proc. Nat. Acad. Sci. U.S.A. *42:* 481–487.

Katsumata, K., and Ozawa, T. (1969). Effect of insulin on mitochondrial respiratory control and oxidative phosphorylation of alloxan diabetic rats. A possible new action of insulin on liver mitochondria. Nagoya J. Med. Sci. *32:* 45–53; CA *71:* (12) 109518s (1969).

Keilin, D. (1925). On cytochrome, a respiratory pigment, common to animals, yeast, and higher plants. Proc. Roy. Soc. (Biol.) *98:* 312–339.

Keilin, D., and Hartree, E. F. (1939). Cytochrome and cytochrome oxidase. Proc. Roy. Soc. (Biol.) *127:* 167–191.

Kendall, E. C. (1929). Thyroxine. New York, The Chemical Catalogue Co., Inc.

Kerppola, W. (1960). Uncoupling of the oxidative phosphorylation with cortisone in liver mitochondria. Endocrinology *67:* 252–263.

Kim, K-H., Blatt, L. M., and Cohen, P. P. (1967). Thyroxine interaction with actinomycin D and possible biological implications. Science *156:* 245–246.

Kim, K-H., and Cohen, P. P. (1968). Actinomycin D inhibition of thyroxine-induced synthesis of carbamyl phosphate synthetase. Biochim. Biophys. Acta *166:* 574–577.

Kimberg, D. V., Goodman, N. F., and Graudusius, R. T. (1969). Effect of glucocorticoid treatment on calcium-activated rat liver mitochondrial adenosine triphosphatase. Endocrinology *84:* 1384–1397.

Kimberg, D. V., Loud, A. V., and Wiener, J. (1968). Cortisone-induced alterations in mitochondrial function and structure. J. Cell Biol. *37:* 63–79.

Klain, G. J. (1963). Alterations of rat-tissue cytochrome *c* levels by a chronic cold exposure. Biochim. Biophys. Acta *74:* 778–780.

Kleiber, Max. (1961). The fire of life: an introduction to animal energetics. New York, John Wiley & Sons, Inc.

Klein, P. D., and Johnson, R. M. (1954). Phosphorus metabolism in unsaturated fatty acid-deficient rats. J. Biol. Chem. *211:* 103–110.

Klingenberg, M. (1963). Morphological and functional aspects of pyridine nucleotide reactions in mitochondria. *In* Chance, B. (ed.): Energy-Linked Functions of Mitochondria. New York, Academic Press, Inc., pp. 121–139.

Klingenberg, M., and Kröger, A. (1967). On the role of ubiquinone in the respiratory chain. *In* Slater, E. C., Kaniuga, Z., and Wojtczak, L. (eds.): Biochemistry of Mitochondria. New York, Academic Press, Inc., pp. 11–27.

Klingenberg, M., and Pfaff, E. (1966). Structural and functional compartmentation in mitochondria. *In* Tager, J. M., Papa, S., Quagliariello, E., Slater, E. C. (eds.): Regulation of Metabolic Processes in Mitochondria. New York, American Elsevier Publishing Company, Inc., pp. 180–201.

Klingenberg, M., and Slenczka, W. (1959). Pyridinnucleotide in Leber-Mitochondrien, Eine Analyse ihrer Redox-Beziehungen. Biochem. Z. *331*: 486–517.

Klitgaard, H. M. (1966). Effect of thyroidectomy on cytochrome *c* concentration of selected rat tissues. Endocrinology *78*: 642–644.

Klotz, I. M. (1967). Energy changes in biochemical reactions. New York, Academic Press, Inc., pp. 1–108.

Kohen, E., Kohen, C., and Jenkins, W. (1966). The influence of microelectrophoretically induced metabolites on pyridine nucleotide reduction in giant tissue culture ascites cells. Exp. Cell Res. *44*: 175–194.

Kohen, E., Kohen, C., Thorell, B., and Akerman, L. (1968). Kinetics of the fluorescence response to micro-electrophoretically introduced metabolites in the single living cell. Biochim. Biophys. Acta *158*: 185–188.

Korner, A. (1960). The effect of the administration of insulin to the hypophysectomized rat on the incorporation of amino acids into liver proteins *in vivo* and in a cell-free system. Biochem. J. *74*: 471–478.

Korr, I. M. (1939). Oxidation-reductions in heterogeneous systems. Sympos. Quant. Biol. *7*: 74–93.

Krall, A. R., Siegel, G. J., and Gozansky, D. (1964). Adrenochrome inhibition of oxidative phosphorylation by rat brain mitochondria. Biochem. Pharmacol. *13*: 1519–1525.

Krebs, H. A., and Kornberg, H. L. (1957). Energy transformations in living matter. A survey. Berlin-Göttingen-Heidelberg Springer-Verlag OHG., Ergebn. Physiol. *49*: 212–298.

Krishna, G., Hynie, S., and Brodie, B. B. (1968). Effects of thyroid hormones on adenyl cyclase in adipose tissue and on free fatty acid mobilization. Proc. Nat. Acad. Sci. U.S.A. *59*: 884–889.

Kunin, A. S., and Krane, S. M. (1965). Inhibition by puromycin of the calcium-mobilizing activity of parathyroid extract. Endocrinology *76*: 343–344.

Lamberg, B. A. (1959). The medical thyroid crisis. Acta Med. Scand. *164*: 479–496.

Lamy, R. M., Rodesch, F. R., and Dumont, J. E. (1967). Action of thyrotropin on thyroid energetic metabolism. VI. Regulation of mitochondrial respiration. Exp. Cell Res. *46*: 518–532.

Lardy, H. A. (1952). The role of phosphate in metabolic control mechanisms. *In* Wolterink, L. F. (ed.): The Biology of Phosphorus. East Lansing, Michigan State University Press, pp. 131–147.

Lardy, H. A. (1954). Effect of thyroid hormones on enzyme systems. *In* The Thyroid. Upton, N.Y., Brookhaven National Laboratory, pp. 90–97 (disc. –101).

Lardy, H. A., and Feldott, G. (1951). Metabolic effects of thyroxine *in vitro*. Ann. N.Y. Acad. Sci. *54*: 636–648.

Lardy, H. A., Johnson, D., and McMurray, W. C. (1958). A survey of toxic antibiotics in respiratory, phosphorylative, and glycolytic systems. Arch. Biochem. *78*: 587–597.

Lardy, H. A., and Lee, Y-P. (1961). Cellular effects of the thyroid hormone in different organs and species. Amer. Zool. *1*: 457.

Lardy, H. A., Lee, Y-P., Takemori, A. (1960). Enzyme responses to thyroid hormones. Ann. N.Y. Acad. Sci. *86*: 506–511.

Lardy, H. A., and Wellman, H. (1952). Oxidative phosphorylations: role of inorganic phosphate and acceptor systems in control of metabolic rates. J. Biol. Chem. *195*: 215–224.

Laszlo, J., Miller, D. S., McCarty, K. S., and Hochstein, P. (1966). Actinomycin D: Inhibition of respiration and glycolysis. Science *151*: 1007–1010.

Lauber, J. K. (1969). *In* The Amateur Scientist (cond. by C. L. Strong). Sci. Amer. *221*: 122–126.

Leblanc, P., Bourdian, M., and Clauser, H. (1968). Oxidophosphorylating properties of pig heart sarcosomes in the presence and absence of insulin. Bull. Soc. Chim. Biol. *50*: 2091–2119; CA *70*: 112008k (1969).

Leblond, C. P., and Eartly, H. (1952). An attempt to produce complete thyroxine deficiency in the rat. Endocrinology *51:* 26–41.

Lee, K-L., and Miller, O. N. (1967). Induction of mitochondrial α-glycerophosphate dehydrogenase by thyroid hormone: Comparison of the euthyroid and the thyroidectomized rat. Arch. Biochem. *120:* 638–645.

Lee, Y-P, and Hsu, H. H-T. (1969). Studies on the effect of thyroid hormones on cytochrome-linked L-α-glycerophosphate dehydrogenase of the mouse. Endocrinology *85:* 251–258.

Lee, Y-P., and Lardy, H. A. (1965). Influence of thyroid hormones on L-α-glycerophosphate dehydrogenases and other dehydrogenases in various organs of the rat. J. Biol. Chem. *240:* 1427–1436.

Lee, Y-P., Takemori, A. E., and Lardy, H. A. (1959). Enhanced oxidation of α-glycerophosphate by mitochondria of thyroid-fed rats. J. Biol. Chem. *234:* 3051–3054.

Lee, N. D., and Williams, R. H. (1954). The intracellular localization of labeled thyroxine and labeled insulin in mammalian liver. Endocrinology *54:* 5–19.

Lee, N. D., and Wiseman, R., Jr. (1969). The significance of the binding of insulin-I[131] to cytostructural elements of rat liver. Endocrinology *65:* 442–450.

Lefebvre, P., and Lelievre, P. (1964). Effects of glucagon on the metabolism of rat mitochondria. Arch. Int. Physiol. *72:* 857–862.

Lehninger, A. L. (1951). Phosphorylation coupled to oxidation of dihydrodiphosphopyridine nucleotide. J. Biol. Chem. *190:* 345–359.

Lehninger, A. L. (1960). Thyroxine and the swelling and contraction cycle in mitochondria. Ann. N.Y. Acad. Sci. *86:* 484–493.

Lehninger, A. L. (1962). Water uptake and extrusion by mitochondria in relation to oxidative phosphorylation. Physiol. Rev. *42:* 467–517.

Lehninger, A. L. (1964). The Mitochondrion. New York, W. A. Benjamin.

Lehninger, A. L. (1965). Bioenergetics: The Molecular Basis of Biological Energy Transformations. New York, W. A. Benjamin.

Lehninger, A. L., and Neubert, D. (1961). Effect of oxytocin, vasopressin, and other disulfide hormones on uptake and extrusion of water by mitochondria. Proc. Nat. Acad. Sci. U.S.A. *47:* 1929–1936.

Lehninger, A. L., and Wadkins, C. L. (1962). Oxidative phosphorylation. Ann. Rev. Biochem. *32:* 47–78.

Lehninger, A. L., Wadkins, C. L., and Remmert, L. T. (1959). Control points in phosphorylating respiration and the action of a mitochondrial respiration-releasing factor. *In* Wolstenholme, G. E. W., and O'Connor, C. M. (eds.): Ciba Foundation Symposium on the Regulation of Cell Metabolism. London, J. & A. Churchill Ltd., pp. 130–145.

Leon-Sotomayor, L., and Bowers, C. Y. (1964). Myxedema Coma. Charles C Thomas, Publisher.

Levell, M. F., and Fourman, P. (1968). Recent advances in the physiology of parathyroid hormone. Ann. Endocr. (Paris) *29:* 576–591.

Lewis, G. N., and Randall, M. (1923). Thermodynamics and the free energy of chemical substances. New York, McGraw-Hill Book Co., Inc.

Lianides, S. P., and Beyer, R. E. (1960a). Thyroid function and oxidative phosphorylation in cold-exposed rats. Nature *188:* 1196–1197.

Lianides, S. P., and Beyer, R. E. (1960b). Oxidative phosphorylation in liver mitochondria prepared from adrenal-demedullated and epinephrine-treated rats. Biochim. Biophys. Acta *44:* 356–357.

Liew, C. C., and Gornall, A. C. (1969). Effects of aldosterone on the incorporation of phosphorus into the nucleotides of heart muscle. Canad. J. Biochem. *47:* 461–466.

Lindberg, O., DePierre, J., Rylander, E., and Afzelius, B. A. (1967). Studies of the mitochondrial energy-transfer system of brown adipose tissue. J. Cell Biol. *45:* 293–310.

Lindsay, R. H., Nakagawa, H., and Cohen, P. P. (1965). Thiouracil-pyrimidine relationships in thyroid and other tissues. Endocrinology *76:* 728–736.

Lindsay, R. H., Romine, C. J., and Song, M. Y. (1968). Effects of 2-thiouracil and related compounds on pyrimidine metabolism. Arch. Biochem. *126:* 812–820.

Linford, J. H. (1966). An Introduction to Energetics, with Applications to Biology. London, Butterworth & Co., Ltd.

Lipmann, F. (1941). Metabolic generation and utilization of phosphate bond energy. Advances Enzym. *1:* 100–162.

Locker, A., Siedek, H., and Spitzy, K. H. (1950). Zur Wirkung von Dinitrokresol und Thiouracil auf den Zellstoffwechsel. Naunyn Schmiedeberg Arch. Exp. Path. Pharmakol. *210:* 281–288.

Loomis, W. F., and Lipmann, F. (1948). Reversible inhibition of the coupling between phosphorylation and oxidation. J. Biol. Chem. *173:* 807–808.

Lowe, C. U., and Lehninger, A. L. (1955). Oxidation and phosphorylation in liver mitochondria lacking "polymerized" ribonucleic acid. J. Biophys. Biochem. Cyt. *1:* 89–92.

Luft, R., Ikkos, D., Palmieri, G., Ernster, L., and Afzelius, B. (1962). A case of severe hypermetabolism of nonthyroid origin with a defect in the maintenance of mitochondrial respiratory control: a correlated clinical, biochemical, and morphological study. J. Clin. Invest. *41:* 1776–1804.

Lundholm, L., Mohme-Lundholm, E., and Svedmyr, N. (1966). Physiological interrelationships: introductory remarks. Pharmacol. Rev. *18:* 255–272.

Lundquist, C-G., and Svanborg, A. (1964). Respiratory activity of mitochondria from human skeletal muscle before and after insulin administration. Acta Med. Scand. *176:* 701–704.

Lyman, C. P., and Chatfield, P. O. (1955). Physiology of hibernation in mammals. Physiol. Rev. *35:* 403–425.

MacDonald, I. R., and Ellis, R. J. (1969). Does cycloheximide inhibit protein synthesis specifically in plant tissues? Nature *222:* 791–792.

Magne, H., Mayer, A., and Plantefol, L. (1931–32). Action pharmaco-dynamique des phénols nitres. Un agent augmentant les oxydations cellulaires. Le Dinitrophenol 1-2-4 (Thermol). Ann. Physiol. Physicochim. Biol. *7:* 269–273; *8:* 1–194.

Magnus-Levy, A. (1895). Über den respiratorischen Gaswechsel unter dem Einfluss der Thyroidea sowie unter verschiedenen pathologischen Zustanden. Berlin. Klin. Wschr. *32:* 650–652.

Mahler, H. R., and Cordes, E. H. (1966). Biological Chemistry. New York, Harper and Row, Publishers.

Makinen, M. W., and Lee, C-P. (1968). Biochemical studies of skeletal muscle mitochondria. I. Microanalysis of cytochrome content, oxidative and phosphorylative activities of mammalian skeletal muscle mitochondria. Arch. Biochem. *126:* 75–82.

Maley, G. F., and Lardy, H. A. (1955). Efficiency of phosphorylation in selected oxidations by mitochondria from normal and thyrotoxic livers. J. Biol. Chem. *215:* 377–388.

Mantegazza, P., and Riva, M. (1965). Amphetamine activity in hypothyroid animals. Atti Accad. Med. Lombard. *20:* 10–15; CA *64:* 2616b (1966).

Marchis-Mouren, G., and Cozzone, A. (1967). Concentration of tritiated actinomycin D in various rat tissues after intraperitoneal injection. Bull. Soc. Chim. Biol. *49:* 569–571; CA *67:* 72156m (1967).

Margaria, R. (1966). Assessment of physical activity in oxidative and anaerobic maximal exercise. Fed. Proc. *25:* 1409–1412.

Margaria, R., Oliva, R. D., DiPrampero, P. E., and Cerretelli, P. (1969). Energy utilization in intermittent exercise of supramaximal intensity. J. Appl. Physiol. *26:* 752–756.

Martin, J. T. (1968). Fulminant hyperthermia. J.A.M.A. *204:* 729.

Martin, I. R., Mecca, C. E., Schiffman, E., and Goldhaber, P. (1965). Alterations in bone metabolism induced by parathyroid extract. *In* Gaillant, P. J., Talmadge, R. V., Budy, A. M. (eds.): The Parathyroid Glands, Ultrastructure, Secretion, and Function. (Noordwijk, 1964). Chicago, U. Chicago Press, pp. 261–272.

Martius, C., and Hess, B. (1951). The mode of action of thyroxine. Arch. Biochem. *33:* 486–487.

Martius, C., and Hess, B. (1952). Über den Wirkungsmechanismus des Schilddrüsenhormons. Arch. Exp Pathol. Pharmakol. *216:* 45–46.

Masoro, E. J. (1966). Effect of cold on metabolic use of lipids. Physiol. Rev. *46:* 67–101.

Masoro, E. J. (1968). Physiological Chemistry of Lipids in Mammals (Physiological Chemistry Series, E. J. Masoro, ed.). Philadelphia, W. B. Saunders Company.

Massey, V., and Veeger, C. (1963). Biological oxidations. Ann. Rev. Biochem. *32:* 579–638.

McArthur, J. W., Rawson, R. W., Means, J. H., and Cope, O. (1947). Thyrotoxic crisis. An analysis of the thirty-six cases seen at the Massachusetts General Hospital during the past twenty-five years. J.A.M.A. *134:* 868–874.

McIver, M. A. (1940). Increased susceptibility to chloroform poisoning produced in the albino rat by injection of crystalline thyroxin. Proc. Soc. Exp. Biol. Med. *45:* 201–206.

McIver, M. A., and Winter, E. A. (1942). Further studies on increased susceptibility to chloroform poisoning produced in the albino rat by injection of crystalline thyroxin. J. Clin. Invest. *21:* 191–196.

McKerns, K. W. (ed.) (1968). Functions of the Adrenal Cortex. Vol. 1. New York, Appleton-Century-Crofts.

Means, J. H., DeGroot, L. T., and Stanbury, J. B. (1963). The thyroid and its diseases. 3rd ed. New York, McGraw-Hill Book Co., Inc.

Means, J. H., and Lerman, J. (1935). Symptomatology of myxedema. Its relation to metabolic levels, time intervals and rations of thyroid. Arch. Intern. Med. *55:* 1–6.

Metzenberg, R. L., Marshall, M., Paik, W. K., and Cohen, P. P. (1961). The synthesis of carbamyl phosphate synthetase in thyroxin-treated tadpoles. J. Biol. Chem. *236:* 162–165.

Minaire, Y., and Chatonnet, J. (1966). Can the glycogenolytic effect alone explain the very high calorigenic action which adrenaline exerts during maximum thermal shock? Arch. Sci. Physiol. *20:* 21–41; CA *65:* 991a (1966).

Mitchell, P. (1966). Chemiosmotic Coupling in Oxidative and Photosynthetic Phosphorylation. Bodmin, England, Glynn Res., Ltd.

Mitchell, P. (1968). Chemiosmotic Coupling and Energy Transduction. Bodmin, England, Glynn Res., Ltd.

Mitchell, P., and Moyle, J. (1969). Estimation of membrane potential and pH difference across the cristae membrane of rat liver mitochondria. European J. Biochem. *7:* 471–484.

Miyazaki, E., Yabu, H., and Takahashi, M. (1962). Increasing effect of caffein on the oxygen consumption of the skeletal muscle. Jap. J. Physiol. *12:* 113–123; CA *57:* 6555g (1962).

Moeschlin, S. (1965). Poisoning, Diagnosis and Treatment. New York, Grune & Stratton, Inc.

Mommaerts, W. F. H. M., Uchida, K., and Seraydarian, K. (1963). Cyclic adenosine 3′,5′-phosphate as a regulator of the contractility of actomyosin. Fed. Proc. *22:* 351.

Moore, K. E. (1966). Amphetamine toxicity in hyperthyroid mice; effects on blood glucose and liver glycogen. Biochem. Pharmacol. *15:* 353–360.

Morehouse, L. E., and Miller, A. T., Jr. (1967). Physiology of Exercise. St. Louis, The C. V. Mosby Co.

Morton, B. E., and Lardy, H. A. (1967). Cellular oxidative phosphorylation. I. Measurement in intact spermatozoa and other cells. Biochemistry *6:* 43–49. II. Measurement in physically modified spermatozoa. Biochemistry *6:* 50–56. III. Measurement in chemically modified cells. Biochemistry *6:* 57–61.

Muldoon, S. M., and Theye, R. A. (1969). Effects of succinylcholine and d-tubo-curarine on oxygen consumption. Anesthesiology *31:* 437–442.

Murray, P. C., and Bisset, S. K. (1963). Effect of 3,3′,5′-triiodothyroacetic acid on the *in vitro* oxygen uptake of leukocytes of hypopituitary patients. Nature *197:* 1209–1210.

Musacchia, X. J., and Saunders, J. F. (eds.) (1969). Depressed metabolism, Proc. First Int. Conf. on Depressed Metabolism, Wash., D.C., 1968. New York, American Elsevier Publishing Company, Inc.

Nakamura, M., Nakatini, M., Koike, M., Torii, S., and Hiramatsu, M. (1961). Swelling of heart and liver mitochondria from magnesium-deficient rats and its reversal. Proc. Soc. Exp. Biol. Med. *108:* 315–319.

Narusawa, J. (1956). 2,4-Dinitrophenol fever. Nippon Yakurigaku Zasshi *52:* (I) 820–827, (II) 828–842, (III) 843–850; CA *52:* 1469d (1958).

Nass, M. M. K. (1966). The circularity of mitochondrial DNA. Proc. Nat. Acad. Sci. U.S.A. *56:* 1215–1222.

Nass, M. M. K. (1969). Mitochondrial DNA: advances, problems, and goals. Science *165:* 26–35.

Nass, M. M. K., and Nass, S. (1963). Intramitochondrial fibers with DNA characteristics. I. Fixation and electron staining reactions. II. Enzymatic and other hydrolytic treatments. J. Cell Biol. *19:* 593–611; 613–629.

Nickerson, J. F., Hill, S. R., McNeil, J. H., and Barker, S. B. (1960). Fatal myxedema with and without coma. Ann. Intern. Med. *53:* 475–493.

Nielsen, P. E., and Ranløv, P. (1964). Myxoedema coma; case reports and a review. Acta. Endocr. *45:* 353–364.

Niemeyer, H., Crane, R. K., Kennedy, E. P., and Lipmann, F. (1951). Observations on respiration and phosphorylation with liver mitochondria of normal, hypo- and hyperthyroid rats. Fed. Proc. *10:* 229.

Noble, R. W. (1969). Relation between allosteric effects and changes in the energy of bonding between molecular subunits. J. Molec. Biol. *39:* 479–491.

Ogata, E. (1962). Experimental studies on the control mechanism of cellular respiration in febrile conditions. Nisshin Igaku *49:* 19–39; CA *57:* 14349j (1962).

Okui, S. (1955). Antipyretic action of quinine hydrochloride. Med. J. Osaka Univ. *6:* 407–416; CA *50:* 9620c (1956).

Opie, L., Lochner, A., Brink, A. J., Homburger, F., and Nixon, C. W. (1964). Oxidative phosphorylation in hereditary myocardiopathy in the Syrian hamster. Lancet *2:* 1213–1214.

Oppenheimer, J. H., Bernstein, G., and Surks, M. I. (1968). Increased thyroxine turnover and thyroidal function after stimulation of hepatocellular binding of thyroxine by phenobarbital. J. Clin. Invest. *47:* 1399–1406.

Packer, L., and Tappel, A. L. (1960). Light scattering changes linked to oxidative phosphorylation in mitochondrial membrane fragments. J. Biol. Chem. *235:* 525–530.

Paik, W. K., and Cohen, P. P. (1960). Biochemical studies on amphibian metamorphosis. I. The effect of thyroxine on protein synthesis in the tadpole. J. Gen. Physiol. *43:* 683–696.

Paik, W. K., and Cohen, P. P. (1961). Biochemical studies on amphibian metamorphosis. II. The effect of thiouracil on thyroxine-stimulated protein synthesis in tadpole liver. J. Biol. Chem. *236:* 531–535.

Paik, W. K., Metzenberg, R. L., and Cohen, P. P. (1961). Biochemical studies on amphibian metamorphosis. III. Metabolism of nucleic acids and nucleotides in tadpole liver during thyroxine-induced metamorphosis. J. Biol. Chem. *236:* 536–541.

Park, J. H., Meriwether, B. P., and Park, C. R. (1956). Effects of adrenochrome on oxidative phosphorylation in liver mitochondria. Fed. Proc. *15:* 141.

Parker, V. H. (1954). The effect of 3,5-dinitro-ortho-cresol on phosphocreatine and the adenosine phosphate compounds of rat tissues. Biochem. J. *57:* 381–386.

Parker, V. H. (1956). *In vivo* inhibition of oxidative phosphorylation of rat liver mitochondria by 2,4-dinitrophenol. Nature *178:* 261.

Parker, V. H. (1965). Uncouplers of rat-liver mitochondrial oxidative phosphorylation. Biochem. J. *97:* 658–662.

Parker, V. H., Barnes, J. M., and Denz, F. A. (1951). Some observations on the toxic properties of 3,5-dinitro-ortho-cresol. Brit. J. Industr. Med. *8:* 226–235.

Parsons, D. F., Williams, G. R., and Chance, B. (1966). Characteristics of isolated and purified preparations of the outer and inner membranes of mitochondria. Ann. N.Y. Acad. Sci. *137:* 643–666.

Patkin, J., and Masoro, E. J. (1960). Effects of cold stress on mitochondrial oxidative phosphorylation. Am. J. Physiol. *199:* 201–202.

Pauling, L. (1961). A molecular theory of general anesthesia. Science *134:* 15–21.

Pauling, L. (1964). The hydrate microcrystal theory of general anesthesia. Anesth. Analg. (Cleveland) *43:* 1–10.

Pedersen, S., Tata, J. R., and Ernster, L. (1963). Ubiquinone (coenzyme Q) and the regulation of basal metabolic rate by thyroid hormones. Biochim. Biophys. Acta *69:* 407–409.

Pemberton, J. D. (1936). Postoperative hyperthyroid reactions. Western J. Surg. *44:* 521–527.

Penniston, J. T., Harris, R. A., Asai, J., and Green, D. E. (1968). The conformational basis of energy transformations in membrane systems. I. Conformational changes in mitochondria. Proc. Nat. Acad. Sci. U.S.A. *59:* 624–631.

Perez, V., Gorosdisch, S., deMartire, J., Nicholson, R., and diPaola, G. (1969). Oral contraceptives: long-term use produces fine structural changes in liver mitochondria. Science *165:* 805–807.

Perkins, R. G. (1919). A study of the munitions intoxications in France. Public Health Rep. *34:* 2335–2374.

Perlgut, L. E., and Wainio, W. W. (1964). Iodohistidine as a possible intermediate in oxidative phosphorylation: its identification and activity. Biochem. Biophys. Res. Commun. *16:* 227–232.

Perlgut, L. E., and Wainio, W. W. (1966). Mitochondrial phosphoriodohistidine. A possible high energy intermediate of oxidative phosphorylation. Biochemistry *5:* 608–618.

Peter, J. B. (1968). Cellular abnormalities in hyperthyroidism. In Hyperthyroidism, The UCLA Interdepartmental Conference. Ann. Intern. Med. *69:* 1016–1035.

Piiper, J., DiPrampero, P. E., and Cerretelli, P. (1968). Oxygen debt and high-energy phosphates in gastrocnemius muscle of the dog. Amer. J. Physiol. *215:* 523–531.

Pittman, C. S., and Barker, S. B. (1959). Inhibition of thyroxine action by 3,3',5'-triiodothyronine. Endocrinology *64:* 466–468.

Poe, M., Gutfreund, H., and Estabrook, R. W. (1967). Kinetic studies of temperature changes and oxygen uptake in a differential calorimeter: the heat of oxidation of NADH and succinate. Arch. Biochem. *122:* 204–211.

Poole, F. E., and Haining, R. B. (1934). Sudden death from dinitrophenol poisoning. J.A.M.A. *102:* 1141–1147.

Potter, V. R. (1958). Possible biochemical mechanisms underlying adaptation to cold. Fed. Proc. *17:* 1060–1063.

Prange, A. J., Jr., and Lipton, M. A. (1966). The influence of thyroid status on the effects and metabolism of pentobarbital and thiopental. Biochem. Pharmacol. *15:* 237–248.

Prusiner, S. B., Cannon, B., Ching, T. M., and Lindberg, O. (1968a). Oxidative metabolism in cells isolated from brown adipose tissue. II. Catecholamine regulated respiratory control. European J. Biochem. *7:* 51–57.

Prusiner, S. B., Cannon, B., and Lindberg, O. (1968b). Oxidative metabolism in cells isolated from brown adipose tissue. I. Catecholamine and fatty acid stimulation of respiration. European J. Biochem. *6:* 15–22.

Prusiner, S. B., Williamson, J. R., Chance, B., and Paddle, B. M. (1968c). Pyridine nucleotide changes during thermogenesis in brown fat tissue *in vivo*. Arch. Biochem. *123:* 368–377.

Pugsley, L. I. (1935). The effect of 2,4-dinitrophenol upon calcium, creatine and creatinine excretion in the rat. Biochem. J. *29:* 2247–2250.

Pullman, M. E., and Schatz, G. (1967). Mitochondrial oxidations and energy coupling. Ann. Rev. Biochem. *36:* 539–610.

Quastel, J. H. (1952). Biochemical aspects of narcosis. Anesth. Analg. (Cleveland) *31:* 151–163.

Racker, E. (1961). Mechanisms of synthesis of adenosine triphosphate. Advances Enzym. *23:* 323–399.

Racker, E. (1965). Mechanisms in Bioenergetics. New York, Academic Press, Inc.

Racker, E. (1968). The membrane of the mitochondrion. Sci. Amer. *218:* 32–39.

Ramasarma, T., Joshi, V. C., Inamdar, A. R., Aithal, H. N., and Krishnaiah, D. V. (1967). Influence of vitamin A, thyroxine, cold-exposure and cholesterol and coenzyme Q feedings on the isoprene metabolism in the rat. Progr. Biochem. Pharmacol. *2:* 56–61.

Raskova, H. (1948). Some pharmacological properties of thiouracil. Arch. Int. Pharmacodyn. *76:* 442–449.

Rasmussen, H., Arnaud, C., and Hawker, C. (1964). Actinomycin D and the response to parathyroid hormone. Science *144:* 1019–1021.

Rawson, R. W. (1964). Physiologic effects of thyroxine in man. Proc. Mayo Clinic. *39:* 637–653.

Rawson, R. W., Rall, J. E., and Sonenberg, M. (1955). The chemistry and physiology of the thyroid. *In* Pincus, G., and Thimann, K. V. (eds.): The Hormones. Vol. 3. New York, Academic Press, Inc., pp. 433–519.

Reid, C., and Kossa, J. (1954). The action of anti-thyroid agents and thyroxine on tissue homogenates. Arch. Biochem. *53:* 321–326.

Relton, S. S., Creighton, R. E., Johnston, W. E., Pelton, D. A., and Conn, A. W. (1966). Hyperpyrexia in association with general anaesthesia in children. Canad. Anaesth. Soc. J. *13:* 419–424.

Revel, M., Hiatt, H. H., and Revel, J-P. (1964). Actinomycin D: an effect on rat liver homogenates unrelated to its action on RNA synthesis. Science *146:* 1311-1313.

Riddle, O., and Smith, G. C. (1935). The effect of temperature on the calorigenic action of dinitrophenol in normal and thyroidectomized pigeons. J. Pharmacol. Exp. Ther. *55:* 173–178.

Riddle, O., Smith, G. C., Bates, R. W., Moran, C. S., and Lahr, E. L. (1936). Action of anterior pituitary hormones on basal metabolism of normal and hypophysectomized pigeons and on a paradoxical influence of temperature. Endocrinology *20:* 1–16.

Ring, G. C. (1942). The importance of the thyroid in maintaining an adequate production of heat during exposure to cold. Amer. J. Physiol. *137:* 582–588.

Ritter, G. (1966). NAD biosynthesis as an early part of androgen action. Molec. Pharmacol. *2:* 125–133.

Roberts, K. E., Firmat, G., Prunier, J., Schwartz, M. K., and Rawson, R. W. (1956). Effect of phosphate in enhancing action of triiodothyronine. Endocrinology *59:* 565–570.

Ronzoni, E., and Ehrenfest, E. (1936). The effect of dinitrophenol on the metabolism of frog muscle. J. Biol. Chem. *115:* 749–768.

Roodyn, D. B. (1968). Mitochondrial biogenesis: germ-free mitochondria. FEBS Letters *1:* 203–205.

Roodyn, D. B., Freeman, K. B., and Tata, J. R. (1965). The stimulation by treatment *in vivo* with triiodothyronine of amino acid incorporation into protein by isolated rat-liver mitochondria. Biochem. J. *94:* 628–641.

Roodyn, D. B., and Wilkie, D. (1968). The Biogenesis of Mitochondria. London, Methuen & Co., Ltd.

Rossini, L., and Salkho. A. (1966). Effect of actinomycin D on the action of thyroid hormone and on compounds stimulating basal metabolism. Boll. Soc. Ital. Biol. Sper. *42:* 1434–1435; CA *66:* 82758c (1967).

Rubner, M. (1894). Die Quelle der thierischen Wärme. Biology *30:* 73–142.

Ryer, R., III, and Murlin, J. R. (1951). The energy metabolism of normal rabbits as influenced by thyrotropic hormone, insulin, thyroxin, and the combination of insulin and thyroxin. Endocrinology *48:* 75–87.

Saidman, L. J., Havard, E. S., and Eger, E. I., II. (1964). Hyperthermia during anesthesia. J.A.M.A. *190:* 1029–1032.

Sallis, J. D., and DeLuca, H. F. (1966). Action of parathyroid hormone on mitochondria. Magnesium- and phosphate-independent respiration. J. Biol. Chem. *241:* 1122–1127.

Sandell, S., Löw, H., and von der Decken, A. (1967). A critical study of amino acid incorporation into protein by isolated liver mitochondria from adult rats. Biochem. J. *104:* 575–584.

Satoskar, R. S., Jathar, V. S., Raut, V. S., and Bhandarkar, S. D. (1965). Evaluation of basal metabolic rate estimated under orally administered pentobarbital and radioiodine tracer test in the diagnosis of thyroid dysfunction. Indian J. Med. Res. *53:* 528–538.

Schaefer, G., and Naegel, L. (1968a). Action of insulin and triiodothyronine on energy-controlled pathways of hydrogen. Biochim. Biophys. Acta *162:* 617–620.

Schaefer, G., and Naegel, L. (1968b). Effect of insulin and triiodothyronine on liver mitochondria *in vivo*. Hoppe Seyler Z. Physiol. Chem. *349:* 1365–1377.

Schatz, G., Haslbrunner, E., and Tuppy, H. (1964). Deoxyribonucleic acid associated with yeast mitochondria. Biochem. Biophys. Res. Commun. *15:* 127–132.

Schottelius, B. A., and Schottelius, D. D. (1968). Basal energy utilization in mammalian skeletal muscle. Proc. Soc. Exp. Biol. Med. *127:* 1228–1231.

Schwartz, A., and Lee, K. S. (1960). Effect of reserpine on heart mitochondria. Nature *188:* 948–949.

Schwartz, A., and Lee, K. S. (1962). Study of heart mitochondria and glycolytic metabolism in experimentally induced cardiac failure. Circ. Res. *10:* 321–332.

Seager, L. D., and Bernstorf, P. W. (1951). Effect of thyroidectomy and propylthiouracil upon the pyrogenic action of Dicumarol. Fed. Proc. *10:* 333–334.

Selenkow, H. A., and Marcus, F. I. (1960). Masked hyperthyroidism and heart disease. Med. Clin. N. Amer. *44:* 1305–1322.

Selkurt, E. E., and Brecher, G. A. (1956). Splanchnic hemodynamics and oxygen utilization during hemorrhagic shock in the dog. Circ. Res. *4:* 693–704.

Sellers, E. A., Reichman, S., and Thomas, N. (1951). Acclimatization to cold: natural and artificial. Amer. J. Physiol. *167:* 644–650.

Sellers, E. A., and You, S. S. (1950). Role of the thyroid in metabolic responses to a cold environment. Amer. J. Physiol. *163:* 81–91.

Serif, G. S., and Seymour, (1961). Antithyroid effects of an iodinated hydroquinone derivative. Proc. Soc. Exp. Biol. Med. *107:* 987–991.

Shambaugh, G. E., III, Balinsky, J. B., and Cohen, P. P. (1969). Synthesis of carbamyl phosphate synthetase in amphibian liver *in vitro*. The effect of thyroxine. J. Biol. Chem. *244:* 5295–5308.

Sheahan, M. M., Wilkinson, J. H., and Maclagan, N. F. (1951). The biological action of substances related to thyroxine. I. The effect of n-alkyl 3,5-diiodo-4-hydroxybenzoates on oxygen consumption in mice. Biochem. J. *48:* 188–192.

Shemano, I., and Nickerson, M. (1959). Mechanisms of thermal responses to pentylenetetrazol. J. Pharmacol. Exp. Ther. *126:* 143–147.

Shemano, I., and Nickerson, M. (1963). Mechanisms of thermal responses to 2,4-dinitrophenol. J. Pharmacol. Exp. Ther. *139:* 88–93.

Shida, H., and Barker, S. B. (1962). Metabolic effects of thyroid hormones and cortisone in the hypophysectomized rat. J. Endocr. *24:* 83–89.

Shorr, E., Barker, S. B., and Malam, M. (1940). Further studies on the restoration of carbohydrate oxidation to diabetic tissue *in vitro* in the absence of insulin. Amer. J. Physiol. *129:* 463–464.

Shorr, E., Richardson, H. B., and Wolff, H. G. (1933–34). Endogenous glycine formation in myopathies and Graves' disease. Proc. Soc. Exp. Biol. Med. *31:* 207–209.

Shy, G. M., and Gonatas, N. K. (1964). Human myopathy with giant abnormal mitochondria. Science *145:* 493–495.

Shy, G. M., Gonatas, N. K., and Perez, M. (1966). Two childhood myopathies with abnormal mitochondria. I. Megaconial myopathy. II. Pleoconial myopathy. Brain *89:* 133–158.

Siedek, H. (1950). Dinitro-*o*-cresol und thiouracil. Wien Klin. Wschr. *62:* 168–170.

Siegel, E., and Tobias, C. A. (1966). End-organ effects of thyroid hormones; subcellular interactions in cultured cells. Science *153:* 763–765.

Simkins, S. (1936). Dinitrophenol and desiccated thyroid in the treatment of obesity. J.A.M.A. *108:* 2110–2117, 2193–2199.

Simon, E. W. (1953). Mechanisms of dinitrophenol toxicity. Biol. Rev. *28:* 453–479.

Simonyi, J., and Szentgyörgyi, D. (1949). The effect of Actedron (phenylisopropylamine) and thyroxine on the body temperature. Arch. Int. Pharmacodyn. *80:* 1–14.

Singer, H. (1901). Über Aspirin. Beitrag zur Kenntniss der Salicylwirkung. Arch. Phys. Ther. *84:* 527–546.

Skobba, T., and Miya, T. S. (1969). Hyperthermic responses and toxicity of chlorpromazine in L-thyroxine sodium-treated rats. Toxic. Appl. Pharmacol. *14:* 176–181.

Skulachev, V. P. (1963). Regulation of the coupling of oxidation and phosphorylation. Proc. Fifth Int. Congr. Biochem. Moscow, 1961. Vol. 5. *In* Slater, E. C. (ed.): Intracellular Respiration. New York, The Macmillan Company, pp. 365–377.

Skulachev, V. P., Maslov, S. P., Sivkova, V. G., Kalinichenko, L. P., and Masliva, G. M. (1963). Cold uncoupling of oxidative phosphorylation in the muscles of albino mice, Biokhimiia *28:* 70–79; CA *59:* 3202c (1963).

Slater, E. C. (1966a). Oxidative Phosphorylation. *In* Florkin, M., and Stotz, E. H. (eds.): Comprehensive Biochemistry. Vol. 16. Biological Oxidations. New York, American Elsevier Publishing Company, Inc., pp. 327–396.

Slater, E. C. (1966b). The respiratory chain and oxidative phosphorylation: some of the unsolved problems. *In* Slater, E. C., Kaniuga, Z., and Wojtczak, L. (eds.): Biochemistry of Mitochondria. New York, Academic Press, Inc., pp. 1–10.

Slater, E. C. (1967). Application of inhibitors and uncouplers for a study of oxidative phosphorylation, *In* Estabrook, R. W., and Pullman, M. E. (eds.): Methods in Enzymology, Vol. X, Oxidation and Phosphorylation. New York, Academic Press, Inc., pp. 48–57.

Slater, E. C., and Hülsmann, W. E. (1959). Control of rate of intracellular respiration. *In* Wolstenholme, G. E. W., and O'Connor, C. M. (eds.): Ciba Found. Symp. Regulation Cell Metab. Boston, Little, Brown and Company, pp. 58–82.

Slebodzinski, A. (1962). Influence of short-lasting changes of ambient temperature on the rate of release of hormonal iodine from the thyroid gland. Acta Med. Pol. *3:* 191–202; CA *57:* 17298e (1962).

Smith, J. A., and DeLuca, H. F. (1964). Structural changes in isolated liver mitochondria of rats during essential fatty acid deficiency. J. Cell. Biol. *21:* 15–26.

Smith, M. J. H. (1963). Salicylates and intermediary metabolism. *In* Dixon, A. S. & J., Smith, M. J. H., Martin, B. K., and Wood, P. M. N. (eds.): Salicylates. Boston Little, Brown and Company, pp. 47–54.

Smith, R. E. (1955–56). Quantitative relations between liver mitochondria metabolism and total body weight in mammals. Ann. N.Y. Acad. Sci. *62:* 405–421.

Smith, R. E. (1960a). Mitochondrial control of oxidative phosphorylation in cold-acclimated rats. Fed. Proc. *19:* 146–151.

Smith, R. E. (1960b). Comparative effects of thyroxin *in vivo* and cold acclimation on metabolic activity of cell fractions from rat liver. Fed. Proc. *19:* 64–70.

Smith, R. E. (1960c). Proc. Int. Sympos. on Cold Acclimation, Buenos Aires, 1959. Fed. Proc. *19:* 1–165.

Smith, R. E. (1964). Thermogenesis and thyroid action. Nature *204:* 1311–1312.

Smith, R. E. (ed.) (1966). Proc. Int. Sympos. on Metabolic Adaptations to Temperature and Altitude, Kyoto, Japan, 1965. Fed. Proc. *26:* 1151–1433.

Smith, R. E., and Fairhurst, A. S. (1958). A mechanism of cellular thermogenesis in cold-adaptation. Proc. Nat. Acad. Sci. U.S.A. *44:* 705–711.

Smith, R. E., and Hoijer, D. J. (1962). Metabolism and cellular function in cold acclimation. Physiol. Rev. *42:* 60–142.

Smith, R. E., and Horwitz, B. A. (1969). Brown fat and thermogenesis. Physiol. Rev. *49:* 330–425.

Smith, R. E., Roberts, J. C., and Hittelman, J. J. (1966). Nonphosphorylating respiration of mitochondria from brown adipose tissue of rats. Science *154:* 653–654.

Snodgrass, P. J., and Piras, M. M. (1966). The effects of halothane on rat liver mitochondria. Biochemistry *5:* 1140–1149.

Sobel, B., Jequier, E., Sjoerdsma, A., and Lovenberg, W. (1966). Effect of catecholamines and adrenergic blocking agents on oxidative phosphorylation in rat heart mitochondria. Circ. Res. *19:* 1050–1061.

Sokoloff, L. (1967). Action of thyroid hormones and cerebral development. Amer. J. Dis. Child. *114:* 498–506.

Sokoloff, L. (1968). Role of mitochondria in the stimulation of protein synthesis by thyroid hormones. *In* San Pietro, A., Lamborg, M. R., Kenney, F. T. (eds.): Regulatory Mechanisms for Protein Synthesis in Mammalian Cells. New York, Academic Press, Inc., pp. 345–367.

Sokoloff, L., Francis, C. M., and Campbell, P. L. (1964). Thyroxine stimulation of amino acid incorporation into protein independent of any action on messenger RNA synthesis. Proc. Nat. Acad. U.S.A. *52:* 728–736.

Sokoloff, L., Roberts, P. A., Januska, M. M., and Kline, J. E. (1968). Mechanisms of stimulation of protein synthesis by thyroid hormones *in vivo*. Proc. Nat. Acad. Sci. U.S.A. *60:* 652–659.

South, F. E., and House, W. A. (1967). Energy metabolism in hibernation. *In* Fisher, K. C. (ed.): Mammalian Hibernation 3, Proc. Int. Sympos. 3rd. ed. New York, American Elsevier Publishing Company, Inc., pp. 305–324 (1965).

Sproull, D. H. (1954). A peripheral action of sodium salicylate. Brit. J. Pharmacol. *9:* 262–264.

Sproull, D. H. (1957). A comparison of sodium salicylate and 2,4-dinitrophenol as metabolic stimulants *in vitro*. Biochem J. *66:* 527–532.

Stancliff, R. C., Williams, M. A., Utsumi, K., and Packer, L. (1969). Essential fatty acid deficiency and mitochondrial function. Arch. Biochem. *131:* 629–642.

Stannard, J. N. (1939). The mechanisms involved in the transfer of oxygen in frog muscle. Sympos. Quant. Biol. *7:* 394–405.

Starr, P. (1964). Further studies on the therapeutic value of sodium dextrothyroxine. Clin. Pharmacol. Ther. *5:* 728–736.

Stassilli, N. R., Kroc, R. L., and Edlin, R. (1960). Selective inhibition of the calorigenic activities of certain thyroxine analogues with chronic thiouracil treatment in rats. Endocrinology *66:* 782–885.

Stein, A. M., Kaplan, N. O., and Ciotti, M. M. (1959). Pyridine nucleotide transhydrogenase. VII. Determination of the reactions with coenzyme analogues in mammalian tissues. J. Biol. Chem. *234:* 979–986.

Stein, O., and Gross, J. (1962). Effect of thyroid hormone on protein biosynthesis by cell-free systems of liver. Proc. Soc. Exp. Biol. Med. *109:* 817–820.

Stephen, C. R. (1967). Fulminant hyperthermia during anesthesia and surgery. J.A.M.A. *202:* 178–182.

Stocker, W. W., Samaha, F. J., and DeGroot, L. J. (1966). Coupled oxidative phosphorylation in muscle of thyrotoxic patients. Program of the 23rd Annual Meeting, Amer. Federation for Clin. Res., May, 1966. J. Clin. Invest. *45:* viii.

Stocker, W. W., Samaha, F. J., and DeGroot, L. J. (1968). Coupled oxidative phosphorylation in muscle of thyrotoxic patients. Amer. J. Med. *44:* 900–909.

Stoner, H. B., Threlfall, C. J., and Green, J. N. (1952). The effect of 3,5-dinitro-orthocresol on the organic phosphates of muscle. Brit. J. Exp. Path. *33:* 398–404.

Strittmatter, P. (1966). Dehydrogenases and flavoproteins. Ann. Rev. Biochem. *35:* 125–156.

Strubelt, O. (1968). The influence of cyclic 3′,5′-AMP and theophylline on oxygen consumption of rats. Biochem. Pharmacol. *17:* 156–158.

Strubelt, O., and Siegers, C. P. (1969). Mechanism of calorigenic action of theophylline and caffeine. Biochem. Pharmacol. *18:* 1207–1220.

Surtshin, A., Cordonnier, J. K., and Lang, S. (1957). Lack of influence of the sympathetic nervous system on the calorigenic response to thyroxine. Amer. J. Physiol. *188:* 503–506.

Suzuki, Y. (1961). Glucagon. V. The effect of glucagon on the cyclophorase system. Tokyo oshi Ika Daig. Z. *76:* 2615–2624; CA *61:* 12280h (1964).

Svedmyr, N. (1966a). Studies on the relationships between some metabolic effects of thyroid hormones and catecholamines in animals and man. Acta Physiol. Scand. *68:* 1–66.

Svedmyr, N. (1966b). The action of triiodothyronine on some effects of adrenaline and noradrenaline in man. Acta Pharmacol. *24:* 203–216.

Swanson, H. E. (1956). Interrelationships between thyroxin and adrenalin in the regulation of oxygen consumption in the albino rat. Endocrinology *59:* 217–225.

Swanson, H. E. (1967). The effect of temperature on the potentiation of adrenalin by thyroxine in the albino rat. Endocrinology *60:* 205–213.

Tainter, M. L. (1934). Low oxygen tensions and temperatures on the actions and toxicity of dinitrophenol. J. Pharmacol. Exp. Ther. *51:* 45–58.

Tainter, M. L., and Cutting, W. C. (1933). Febrile, respiratory and some other actions of dinitrophenol. J. Pharmacol. Exp. Ther. *48:* 410–429.

Tamada, T., Kajihara, A., Onaya, K., Kobayashi, I., Takemura, Y., and Shichijo, K. (1965). Acute stimulatory effect of cold on thyroid activity and its mechanism in guinea pig. Endocrinology *77:* 968–976.

Tapley, D. F. (1962). The mechanism of action of thyroid hormones. Amer. Zool. *2:* 373–378.

Tapley, D. F. (1964). Mode and site of action of thyroxine. Proc. Mayo Clin. *39:* 626–636.

Tapley, D. F., and Cooper, C. (1956). Effect of thyroxine on the swelling of mitochondria isolated from various tissues of the rat. Nature *178:* 1119.

Tashjian, A. H., Ontjes, D. A., and Goodfriend, T. L. (1964). Mechanism of parathyroid hormone action: effects of actinomycin D on hormone-stimulated ion movements *in vivo* and *in vitro*. Biochem. Biophys. Res. Commun. *16:* 209–215.

Tata, J. R. (1963). Inhibition of the biological action of thyroid hormones by actinomycin D and puromycin. Nature *197:* 1167–1168.

Tata, J. R. (1964). Basal metabolic rate and thyroid hormones. Advances Metab. Dis. *1:* 153–189.

Tata, J. R. (1966). The regulation of mitochondrial structure and function by thyroid hormones under physiological conditions. *In* Tager, J. M., Papa, S., Quagliariello, E., Slater, E. C. (eds.): Regulation of Metabolic Processes in Mitochondria. New York, American Elsevier Publishing Company, Inc., pp. 489–507.

Tata, J. R. (1967). The formation and distribution of ribosomes during hormone-induced growth and development. Biochem. J. *104:* 1–16.

Tata, J. R. (1968). Hormonal regulation of growth and protein synthesis. Nature *219:* 331–337.

Tata, J. R. (1970). Regulation of protein synthesis by growth and developmental hormones. *In* Litwack, G. (ed.): Biochemical Actions of Hormones. Vol. 1. New York, Academic Press, Inc., pp. 89–133.

Tata, J. R., Ernster, L., and Lindberg, O. (1962). Control of basal metabolic rate by thyroid hormones and cellular function. Nature *193:* 1058–1060.

Tata, J. R., Ernster, L., Lindberg, O., Arrhenius, E., Pedersen, S., and Hedman, R. (1963). Action of thyroid hormones at the cell level. Biochem. J. *86:* 408–428.

Tata, J. R., and Widnell, C. C. (1966). Ribonucleic acid synthesis during the early action of thyroid hormones. Biochem. J. *98:* 604–620.

Thiers, R. E., Reynolds, E. S., and Vallee, B. L. (1960). The effect of carbon tetrachloride poisoning on subcellular metal distribution in rat liver. J. Biol. Chem. *235:* 2130–2133.

Thiers, R. E., and Vallee, B. L. (1957). Distribution of metals in subcellular fractions of rat liver. J. Biol. Chem. *226:* 911–920.

Thomson, J. F., Smith, D. E., Nance, S. L., and Habeck, D. A. (1968). Some metabolic characteristics of brown fat, with particular reference to the mitochondria. Comp. Biochem. Physiol. *25:* 783–804.

Thorn, G. W., and Eder, H. A. (1946). Studies on chronic thyrotoxic myopathy. Amer. J. Med. *1:* 583–601.

Thorn, W., Gercken, G., and Hurter, P. (1968). Function, substrate supply, and metabolite content of rabbit heart perfused *in situ*. Amer. J. physiol. *214:* 139–145.

Thut, W. H., and Davenport, H. T. (1966). Hyperpyrexia associated with succinylcholine induced muscle rigidity: a case report. Canad. Anaesth. Soc. J. *13:* 425–428.

Tissières, A. (1946). Les relations de la fonction des glandes thyroide et surrenales avec le taux musculaire du cytochrome C chez le rat. Arch. Int. Physiol. *54:* 305–313.

Tobias, J. M. (1960). Further studies on the nature of the excitable system in nerve. J. Gen. Physiol. *43:* 57–61.

Tobin, R. B., and Slater, E. C. (1965). The effect of oligomycin on the respiration of tissue slices. Biochim. Biophys. Acta *105:* 214–220.

Tonoue, T., and Matsumoto, K. (1961). Thyroxine uptake by liver mitochondria of the cold-exposed rat. Endocrinology *69:* 466–472.

Trey, C., Lipworth, L., Chalmers, T. C., Davidson, C. S., Gottlieb, L. S., Popper, H., and Saunders, S. J. (1968). Fulminant hepatic failure. Presumable contribution of halothane. New Eng. J. Med. *279:* 798–801.

Tucker, V. A. (1969). The energetics of bird flight. Sci. Amer. *200:* 70–78.

Turner, M. L. (1946). The effect of thyroxine and dinitrophenol on the thermal responses to cold. Endocrinology *38:* 263–269.

Vallee, B. L., and Hoch, F. L. (1961). Pyridine nucleotide dependent metallodehydrogenases. Ergebn. Physiol. *51:* 52–97.

van Uytvanck, P. (1931). Recherches sur l'action hyperthermisante du dinitro-alpha-naphthol chez le pigeon. Arch. Int. Pharmacodyn. *41:* 160–212.

van Wijngaarden, G. K., Bethlem, J., Meijer, A. E. F. H., Hülsmann, W. C., and Feltkamp, C. A. (1967). Skeletal muscle disease with abnormal mitochondria. Brain *90:* 577–592.

Vester, J. W., and Stadie, W. C. (1957). Oxidative phosphorylation by hepatic mitochondria from the diabetic cat. J. Biol. Chem. *227:* 669–676.

Viguera, M. G., and Conn, A. W. (1967). An investigation of general anaesthesia and hyperpyrexia in chickens. Canad. Anaesth. Soc. J. *14:* 193–196.

Vitale, J. J., Nakamura, M., and Hegsted, D. M. (1957b). Effect of magnesium deficiency on oxidative phosphorylation. J. Biol. Chem. *228:* 573–576.

Vitale, J. J., Hegsted, D. M., Nakamura, M., and Connors, P. (1957a). The effect of thyroxine on magnesium requirement. J. Biol. Chem. *226:* 597–601.

Volfin, P., and Sanadi, D. R. (1968a). Rapidly labeled mitochondrial protein fraction: early thyroxine effect. J.A.S. Cell Biol. (Abstracts) *39:* 138a–139a.

Volfin, P., and Sanadi, D. R. (1968b). Early effect of thyroxine on a rapidly labeled mitochondrial protein; *In* San Pietro, A., Lamborg, M. R., Kenney, F. T. (eds.): Regulatory Mechanisms for Protein Synthesis in Mammalian Cells. New York, Academic Press, Inc., pp. 259–261.

von Euler, U. S. (1932). Action stimulante du dinitro-α-naphthol, du bleu de méthylène et des substances apparentées sur les échanges respiratoires *in vivo* et *in vitro*. Arch. Int. Pharmacodyn. *43:* 67–85.

von Euler, U. S. (1933). Influence du dinitro-α-naphthol sodique et du bleu de méthylène sur la consommation d'oxygène *in vitro* au muscle de divers animaux. Arch. Int. Pharmacodyn. *44:* 464–479.

Waldstein, S. S., Slodki, S. J., Kaganiec, I., and Bronsky, D. (1960). A clinical study of thyroid storm. Ann. Intern. Med. *52:* 626–642.

Wang, E. (1946). Creatine metabolism and endocrine regulation. Acta Med. Scand. (Suppl.) *169:* 1–81.

Wang, J. K., Moffitt, E. A., and Rosevear, J. W. (1969). Oxidative phosphorylation in acute hyperthermia. Anesthesiology *30:* 439–442.

Warburg, O. (1930). The Metabolism of Tumors, trans. by F. Dickens. New York, Richard Smith, Inc., pp. 1–327.

Warburg, O., and Christian, W. (1936). Pyridin, der wasserstoffübertragende Bestandteil von Gärungsfermenten. Biochem. Z. *287:* 291–328.

Wayne, E. J. (1960). Clinical and metabolic studies in thyroid disease. Brit. Med. J. *i:* 78–90.

Weinbach, E. C., and Garbus, J. (1966). Restoration by albumin of oxidative phosphorylation and related reactions. J. Biol. Chem. *241:* 169–175.

Weinbach, E. C., and Garbus, J. (1968). Structural changes in mitochondria induced by uncoupling reagents; the response to proteolytic enzymes. Biochem. J. *106:* 711–717.

Weinbach, E. C., and Garbus, J. (1969). Mechanism of action of reagents that uncouple oxidative phosphorylation. Nature *221:* 1016–1018.

Weiss, W. P., and Sokoloff, L. (1963). Reversal of thyroxine induced hypermetabolism by puromycin. Science *140:* 1324–1326.

Wesson, L. G., and Burr, G. O. (1931). The metabolic rate and respiratory quotients of rats on a fat-deficient diet. J. Biol. Chem. *91:* 525–539.

Whitehead, R. G., and Weidmann, S. M. (1959). The effect of parathormone on the uptake of ^{32}P into adenosine triphosphate and bone salt in kittens. Biochem. J. *71:* 312–318.

Whitehouse, M. W. (1964). Biochemical properties of anti-inflammatory drugs. III. Uncoupling of oxidative phosphorylation in a connective tissue (cartilage) and liver mitochondria by salicylate analogs; relation of structure to activity. Biochem. Pharmacol. *13:* 319–336.

Widnell, C. C., and Tata, J. R. (1963). Stimulation of nuclear RNA polymerase during the latent period of action of thyroid hormones. Biochim. Biophys. Acta *72:* 506–508.

Widnell, C. C., and Tata, J. R. (1966). Additive effects of thyroid hormone, growth hormone and testosterone on deoxyribonucleic acid-dependent ribonucleic acid polymerase in rat-liver nuclei. Biochem. J. *98:* 621–629.

Williams, R. J. P. (1969). Electron transfer and energy conservation. Curr. Topics in Bioenergetics *3:* 79–156.

Williamson, J. R. (1964). Metabolic effects of epinephrine in the isolated, perfused rat heart. J. Biol. Chem. *230:* 2721–2729.

Williamson, J. R. (1966). Metabolic effects of epinephrine in the perfused rat heart. II. Control steps of glucose and glycogen metabolism. Molec. Pharmacol. *2:* 206–220.

Williamson, J. R., and Jamieson, D. (1966). Metabolic effects of epinephrine in the perfused rat heart. I. Comparison of intracellular redox states, tissue pO_2, and force of contraction. Molec. Pharmacol. *2:* 191–205.

Wilson, O. (1966). Field study of the effect of cold exposure and increased muscular activity upon metabolic rate and thyroid function in man. Fed. Proc. *25:* 1357–1362.

Wilson, R. D., Dent, T. F., Traber, D. L., McCoy, N. R., and Allen, C. R. (1967). Malignant hyperpyrexia with anesthesia. J.A.M.A. *202:* 183–186.

Wilson, R. D., Nichols, R. J., Dent, T. E., and Allen, C. R. (1966). Disturbances of the oxidative-phosphorylation mechanism as a possible etiological factor in sudden unexplained hyperthermia occurring during anesthesia. Anesthesiology *27:* 231–232.

Winter, J. E., and Barbour, H. G. (1927–28). Influence of magnesium salts upon the toxicity and antipyretic action of the salicylates. Proc. Soc. Exp. Biol. Med. *25:* 587–590.

Wiswell, J. G. (1961). Some effects of magnesium loading in patients with thyroid disorders. J. Clin. Endocr. *21:* 31–38.

Wolff, E. C., and Wolff, J. (1964). The mechanism of action of the thyroid hormones. *In* Pitt-Rivers, R., and Trotter, W. R. (eds.): The Thyroid Gland. Vol. 1. London, Butterworth & Co., Ltd., pp. 237–282.

Wyman, J. (1964). Linked functions and reciprocal effects in hemoglobin. A second look. Advances Protein Chem. *19:* 223–286.

Wynn, J., and Fore, W. (1965). The effect of hindered phenols on mitochondrial oxidative phosphorylation. J. Biol. Chem. *240:* 1766–1771.

Yatvin, M. B., Wannemacher, R. W., Jr., and Banks, W. L., Jr. (1964). Effects of thiouracil and of thyroidectomy on liver protein metabolism. Endocrinology *74:* 878–884.

Zimny, M. L., and Taylor, S. (1965). Cardiac metabolism in the hypothermic ground squirrel and rat. Amer. J. Physiol. *208:* 1247–1252.

INDEX

Page numbers in *italics* refer to illustrations.

209